Advanced Electrochemical Biosensors

Advanced Electrochemical Biosensors

Editor

Tae Hyun Kim

MDPI • Basel • Beijing • Wuhan • Barcelona • Belgrade • Manchester • Tokyo • Cluj • Tianjin

Editor
Tae Hyun Kim
Department of Chemistry, Soonchunhyang University
Korea

Editorial Office
MDPI
St. Alban-Anlage 66
4052 Basel, Switzerland

This is a reprint of articles from the Special Issue published online in the open access journal *Applied Sciences* (ISSN 2076-3417) (available at: https://www.mdpi.com/journal/applsci/special_issues/advanced_electrochemical_biosensors).

For citation purposes, cite each article independently as indicated on the article page online and as indicated below:

LastName, A.A.; LastName, B.B.; LastName, C.C. Article Title. *Journal Name* **Year**, *Volume Number*, Page Range.

ISBN 978-3-0365-1116-0 (Hbk)
ISBN 978-3-0365-1117-7 (PDF)

© 2021 by the authors. Articles in this book are Open Access and distributed under the Creative Commons Attribution (CC BY) license, which allows users to download, copy and build upon published articles, as long as the author and publisher are properly credited, which ensures maximum dissemination and a wider impact of our publications.

The book as a whole is distributed by MDPI under the terms and conditions of the Creative Commons license CC BY-NC-ND.

Contents

About the Editor vii

Preface to "Advanced Electrochemical Biosensors" ix

Tae Hyun Kim
Toward Emerging Innovations in Electrochemical Biosensing Technology
Reprinted from: *Applied Sciences* **2021**, *11*, 2461, doi:10.3390/app11062461 1

Boris Lakard
Electrochemical Biosensors Based on Conducting Polymers: A Review
Reprinted from: *Applied Sciences* **2020**, *10*, 6614, doi:10.3390/app10186614 5

Maroua Hamami, Noureddine Raouafi and Hafsa Korri-Youssoufi
Self-Assembled MoS_2/ssDNA Nanostructures for the Capacitive Aptasensing of Acetamiprid Insecticide
Reprinted from: *Applied Sciences* **2021**, *11*, 1382, doi:10.3390/app11041382 29

Martin Lundblad, David A. Price, Jason J. Burmeister, Jorge E. Quintero, Peter Huettl, Francois Pomerleau, Nancy R. Zahniser and Greg A. Gerhardt
Tonic and Phasic Amperometric Monitoring of Dopamine Using Microelectrode Arrays in Rat Striatum
Reprinted from: *Applied Sciences* **2020**, *10*, 6449, doi:10.3390/app10186449 43

Brigitte Bruijns, Roald Tiggelaar and Han Gardeniers
A Microfluidic Approach for Biosensing DNA within Forensics
Reprinted from: *Applied Sciences* **2020**, *10*, 7067, doi:10.3390/app10207067 59

Sophia Karastogianni and Stella Girousi
Electrochemical (Bio)Sensing of Maple Syrup Urine Disease Biomarkers Pointing to Early Diagnosis: A Review
Reprinted from: *Applied Sciences* **2020**, *10*, 7023, doi:10.3390/app10207023 75

About the Editor

Tae Hyun Kim is a Professor in the Department of Chemistry at Soonchunhyang University (Republic of Korea). He received his PhD in chemistry at Seoul National University (Republic of Korea) in 2007. After that, he worked in the Department of Physics at the same university from 2007 to 2009, and in the Department of Bioengineering at University of California, Berkeley (USA) from 2009 to 2010 as a postdoctoral researcher. His current research is directed toward the pursuit of developing bio/chemical sensors based on nano-electrochemical systems such as carbon-nanotube-, graphene-, and nanoparticle-derived sensors.

Preface to "Advanced Electrochemical Biosensors"

Electrochemical sensors possess various advantages over conventional sensors, such as high sensitivity and selectivity, simple instrumentation, portability, outstanding compatibility, short analysis time, and low cost. Thus, various types of sensors based on electrochemical techniques have been developed for the detection of chemically, biologically, and environmentally important analytes. With recent developments in advanced material science and electronic technology such as signal processing and front-end electronic systems, electrochemical sensing methods are comprising a very wide range of analytical possibilities. Among these, electrochemical biosensors have attracted significant interest for the detection of biochemical compounds such as biological proteins, nucleotides, and even tissues due to their practical applications in health care, early diagnosis, and environmental monitoring. Thus, this Special Issue serves the need to promote exploratory research and development in emerging electrochemical biosensor technologies while aiming to present the latest technological and methodological developments in this interdisciplinary field. We invite contributions on topics that include but are not limited to various state-of-the-art electrochemical biosensing technologies.

Tae Hyun Kim
Editor

Editorial

Toward Emerging Innovations in Electrochemical Biosensing Technology

Tae Hyun Kim [1,2]

1. Department of Chemistry, Graduate School, Soonchunhyang University, Asan 31538, Korea; thkim@sch.ac.kr
2. Department of ICT Environmental Health System, Graduate School, Soonchunhyang University, Asan 31538, Korea

Abstract: With the progress of nanoscience and biotechnology, advanced electrochemical biosensors have been widely investigated for various application fields. Such electrochemical sensors are well suited to miniaturization and integration for portable devices and parallel processing chips. Therefore, advanced electrochemical biosensors can open a new era in health care, drug discovery, and environmental monitoring. This Special Issue serves the need to promote exploratory research and development on emerging electrochemical biosensor technologies while aiming to reflect on the current state of research in this emerging field.

Keywords: integrated biosensors; lab-on-a-chip; immunosensors; aptasensors; medical diagnostics; nanomaterials; advanced sensing platforms; environmental monitoring

Citation: Kim, T.H. Toward Emerging Innovations in Electrochemical Biosensing Technology. *Appl. Sci.* **2021**, *11*, 2461. https://doi.org/10.3390/app11062461

Received: 5 March 2021
Accepted: 9 March 2021
Published: 10 March 2021

Publisher's Note: MDPI stays neutral with regard to jurisdictional claims in published maps and institutional affiliations.

Copyright: © 2021 by the author. Licensee MDPI, Basel, Switzerland. This article is an open access article distributed under the terms and conditions of the Creative Commons Attribution (CC BY) license (https://creativecommons.org/licenses/by/4.0/).

The last decade has been marked by the identification of the rapidly emerging innovations in nanosystems and biotechnology [1–3]. These innovations have accelerated the creation of new electrochemical biosensors with remarkable improvements in sensitivity, selectivity, accuracy, and multiplexing capacity, along with significant size reductions [4]. Electrochemical biosensors consist of three parts: a sensitive biocomponent that recognizes the analyte, an electrochemical signal transducer or detector component that transforms the recognition into a measurable electrochemical signal, and an amplification and reader device (Figure 1). With a rich inventory of advanced material science and electronic technology such as signal processing and front-end electronic systems, researchers have overcome the limitations of conventional sensors, such as low sensitivity and lack of availability of miniaturization and integration to parallel processing chips. This Special Issue is dedicated to original results, achievements, and reviews by active researchers working on current state-of-the-art research of electrochemical biosensors. In this editorial, we intend to introduce the topic of the Special Issue, briefly describe each of the contributions that make up this Special Issue, and provide some perspectives on the future development of electrochemical biosensors.

One of the ongoing issues met by biosensing devices is the immobilization process used to intimately connect the bio-specific element onto the transducer without the loss of selectivity and sensitivity. Boris Lakard summarized the latest efforts to develop efficient immobilization approaches for biorecognition elements by entirely maintaining their biological activity, through the utilization of conducting polymers [5]. Conducting polymers have been employed in developing high-performance electrochemical biosensors owing to their advantages such as their charge transport properties and chemical versatility. In particular, conducting polymers can be easily modified and functionalized, which enables immobilization of biorecognition molecules efficiently without the inactivation of their biological properties. In his review, Boris Lakard showed the recent progress in the application of conducting polymers in the recognition of biotargets leading to the development of enzymatic biosensors, immunosensors, DNA biosensors, and whole-cell biosensors with cost-effectiveness and high sensitivity. The improvement in the sensitivity can also be produced by utilizing various emerging nanomaterials including 0D (quantum dots), 1D (nanowires, nanotubes), and 2D (thin films, few layers) materials. Since the discovery of graphene by Andre Geim and Konstantin Novoselov, 2D layered materials, such as graphene, black phosphorous, transition metal dichalcogenides (TMDCs), MXene, and hexagonal boron nitride, have drawn extensive attention due to their excellent properties and various application possibilities. As one of those examples, Hamami et al. reported on an electrochemical aptasensor based on molybdenum disulfide (MoS_2) nanosheets, one of TMDCs in this Special Issue [6]. They demonstrated that the proposed MoS_2-based sensor exhibits rapid detection of acetamiprid insecticide with high sensitivity.

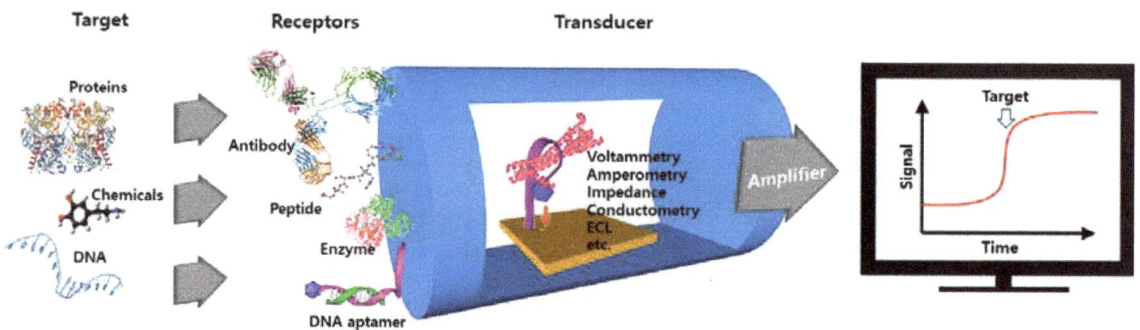

Figure 1. Illustration depicting electrochemical biosensors and their signal transduction principle.

Electrochemical sensing devices are highly suitable for miniaturization since they can be made by conventional microfabrication methods and their analytical performance is maintained even in reduced size. This enables the electrochemical devices to be implemented in various type of microsystems. In this context, two papers aimed at the utilization of electrochemical sensing techniques in microelectrode arrays and microfluidic devices. Lundblad et al. demonstrated a novel microelectrode array approach for monitoring the release of dopamine in the rat striatum [7]. The proposed approach showed highly sensitive detection of tonic (resting) and phasic release of dopamine with subsecond temporal resolution in vivo. Meanwhile, Bruijns et al. reported a microfluidic approach for DNA analysis in forensics [8]. They claimed that the microfluidic devices improve the chain of custody, reduce the contamination risk, and offer fast analysis, enabling them to be used at crime scenes. Indeed, they showed that cyclic olefin copolymer (COC)-based microfluidic device chips could be employed for real-time monitoring of DNA amplification down to 0.01 ng/μL.

Taking advantage of their high sensitivity and selectivity, simple instrumentation, portability, outstanding compatibility, short analysis time, and low cost, advanced electrochemical biosensors have been widely used for various applications in the medical and healthcare sector, the food industry, and environmental monitoring. As a medical application of electrochemical sensors, early diagnosis of metabolic errors was introduced, and the related research was reviewed by Karastogianni et al. [9]. Their review summarized various electrochemical biosensors and point-of-care devices for the detection of branched-chain amino acids which are biomarkers for maple syrup urine disease, an inherited metabolic disorder in which the body cannot process certain amino acids properly.

To summarize, with the continuous progress of nanobiotechnology and the rapid development of electronic devices, innovative frontiers on electrochemical sensors have been launched, enabling prompt utilization of the sensors in various applications, such as medical diagnosis, drug discovery, environmental monitoring, the food industry, and light and heavy chemical industries. A forthcoming advancement in electrochemical biosensors may offer flexible and smooth integration into next-generation ICT systems, making everyday life smarter and easier.

Acknowledgments: This work was conducted with the support of the Korea Environment Industry & Technology Institute (KEITI), through its Ecological Imitation-based Environmental Pollution Management Technology Development Project, and funded by the Korea Ministry of Environment (MOE) (2019002800001). This work was also supported by the Soonchunhyang University Research fund.

Conflicts of Interest: The author declares no conflict of interest.

References

1. Newberry, D. *Nanotechnology Past and Present: Leading to Science, Engineering, and Technology*; Synthesis lectures on engineering, science, and technology; Morgan & Claypool: San Rafael, CA, USA, 2020; ISBN 978-1-68173-861-1.
2. Webster, T.J.; Yazici, H. (Eds.) *Biomedical Nanomaterials: From Design to Implementation*; Healthcare technologies series; The Institution of Engineering and Technology: London, UK, 2016; ISBN 978-1-84919-964-3.
3. Salar, R.K. *Biotechnology: Prospects and Applications*; Springer India: New Delhi, India, 2013; ISBN 978-81-322-1683-4.
4. Zhu, C.; Yang, G.; Li, H.; Du, D.; Lin, Y. Electrochemical sensors and biosensors based on nanomaterials and nanostructures. *Anal. Chem.* **2015**, *87*, 230–249. [CrossRef] [PubMed]
5. Lakard, B. Electrochemical biosensors based on conducting polymers: A review. *Appl. Sci.* **2020**, *10*, 6614. [CrossRef]
6. Hamami, M.; Raouafi, N.; Korri-Youssoufi, H. Self-assembled MoS2/SsDNA nanostructures for the capacitive aptasensing of acetamiprid insecticide. *Appl. Sci.* **2021**, *11*, 1382. [CrossRef]
7. Lundblad, M.; Price, D.A.; Burmeister, J.J.; Quintero, J.E.; Huettl, P.; Pomerleau, F.; Zahniser, N.R.; Gerhardt, G.A. Tonic and phasic amperometric monitoring of dopamine using microelectrode arrays in rat striatum. *Appl. Sci.* **2020**, *10*, 6449. [CrossRef]
8. Bruijns, B.; Tiggelaar, R.; Gardeniers, H. A microfluidic approach for biosensing DNA within forensics. *Appl. Sci.* **2020**, *10*, 7067. [CrossRef]
9. Karastogianni, S.; Girousi, S. Electrochemical (bio)sensing of maple syrup urine disease biomarkers pointing to early diagnosis: A review. *Appl. Sci.* **2020**, *10*, 7023. [CrossRef]

Review

Electrochemical Biosensors Based on Conducting Polymers: A Review

Boris Lakard

UFR Sciences et Techniques, University Bourgogne Franche-Comté, Institut Utinam UMR CNRS 6213, 16 Route de Gray, 25030 Besançon, France; boris.lakard@univ-fcomte.fr; Tel.: +33-381-662-046

Received: 31 August 2020; Accepted: 20 September 2020; Published: 22 September 2020

Abstract: Conducting polymers are an important class of functional materials that has been widely applied to fabricate electrochemical biosensors, because of their interesting and tunable chemical, electrical, and structural properties. Conducting polymers can also be designed through chemical grafting of functional groups, nanostructured, or associated with other functional materials such as nanoparticles to provide tremendous improvements in sensitivity, selectivity, stability and reproducibility of the biosensor's response to a variety of bioanalytes. Such biosensors are expected to play a growing and significant role in delivering the diagnostic information and therapy monitoring since they have advantages including their low cost and low detection limit. Therefore, this article starts with the description of electroanalytical methods (potentiometry, amperometry, conductometry, voltammetry, impedometry) used in electrochemical biosensors, and continues with a review of the recent advances in the application of conducting polymers in the recognition of bioanalytes leading to the development of enzyme based biosensors, immunosensors, DNA biosensors, and whole-cell biosensors.

Keywords: conducting polymers; biosensors; electrochemistry; bioanalyte

1. Introduction

Conducting polymers have attracted much interest since Shirakawa et al. demonstrated in 1977 that halogen doping of polyacetylene strongly increased its conductivity [1]. Thanks to this revolutionary research, Shirakawa, MacDiarmid, and Heeger were awarded the Nobel Prize in Chemistry in 2000, and opened the way to the development of other conducting polymers combining properties of organic polymers and electronic properties of semiconductors. Another major breakthrough in this field was achieved by Diaz et al., who reported the electrodeposition of highly conductive, stable and processable polypyrrole films [2–4]. Following these pioneering studies, numerous conducting polymers have been prepared and used in various applications, such as polyacetylene, polypyrrole (PPy), polyaniline (PANI), polycarbazole, polythiophene (PTh), poly(3,4-ethylenedioxythiophene) (PEDOT), polyphenylene, poly(phenylene vinylene), and polyfluorene (Table 1). All these organic polymers are characterized by alternating single (σ) and double (π) bonds and by the presence of π electrons delocalized across their entire conjugated structure, thus resulting in polymers which can be easily oxidized or reduced [5]. This doping, that can be performed upon oxidation (p-doping) or reduction (n-doping), increases significantly the conductivity of the polymers since this conductivity can vary from less than 10^{-6} S/cm in the neutral state [5] to more than 10^5 S/cm in the doped state [6,7]. The conductivity of the polymers is also dependent on a number of factors including the nature and concentration of the dopant [8–10], temperature [11–13], swelling/deswelling [14], polymer morphology [8], pH and applied potential [15], and polymer chain length [16]. For most heterocyclic polymers, such as PPy [17] or PTh [18], the mechanism of conduction corresponds to a p-doping and starts with the removal of one electron from the initial monomer leading to the formation of an unstable radical cation (named polaron). Then, a second electron is removed from another monomer or from an oligomer, leading to the formation of a dication (named bipolaron) [19]. Under an applied electric field, these polarons and bipolarons serve as charge carriers which are delocalized over the polymer chains and their movement along polymer chains produces electronic conductivity [20].

Table 1. Chemical structures of some common conducting polymers: (a) polyacetylene, (b) polypyrrole, (c) polyaniline, (d) polycarbazole, (e) polythiophenes, (f) poly(3,4-ethylenedioxythiophene), (g) polyphenylenes, (h) poly(phenylene vinylene), (i) polyfluorene.

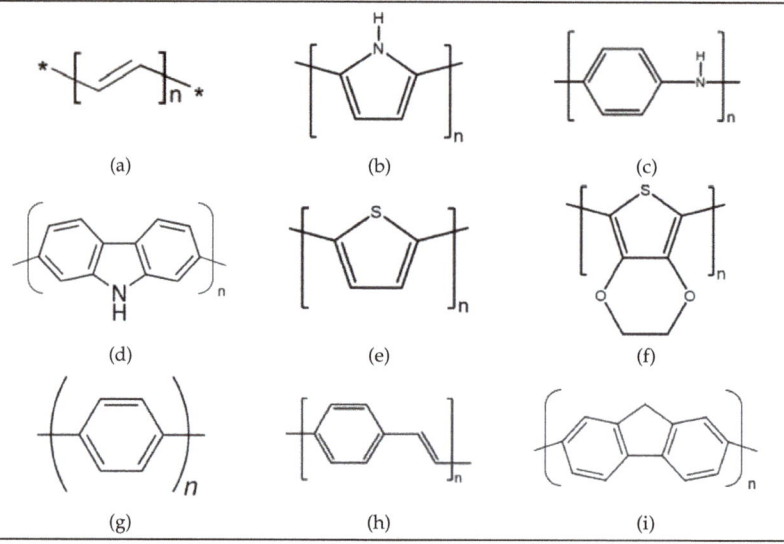

Conducting polymers have become an important class of materials since they combine some useful properties of organic polymers (such as strength, plasticity, flexibility, toughness or elasticity) with unusual electronic [5], optical [21,22] and thermoelectric [23,24] properties due to the charge mobility along the π electron polymer chains. These unique properties explain the use of conducting polymers in a wide variety of applications including energy storage with rechargeable batteries [25,26] and supercapacitors [27,28], photovoltaics with solar cells [29–32], light-emitting diodes [33,34], electrocatalysis [35], anti-corrosion [36,37] or electrochromic applications such as electrochromic displays [38,39] or rearview mirrors and smart windows [40,41].

2. Preparation of Sensitive Materials

2.1. Preparation of Conducting Polymers

Although it is possible to prepare conducting polymers using gas phase techniques such as CVD [42] or plasma polymerization [43,44], conducting polymers are mostly prepared via chemical or electrochemical oxidative polymerization even if it is sometimes possible to use non-oxidative chemical polymerization methods such as Grignard metathesis [45] or dehydrobrominative polycondensation [46]. In traditional chemical oxidative polymerization [47], the synthesis of polymers can be done under harsh oxidative conditions with the use of oxidants such as $K_2Cr_2O_7$, $KMnO_4$, $K_2S_2O_8$, KIO_3 and $FeCl_3$ [48], or under mild conditions by using, for example, the catalytic action of redox enzymes to produce hydrogen peroxide that initiates the polymerization [49], or less frequently at the liquid/air interface [50]. However, the electrochemical oxidative polymerization is the most frequently used method, mainly because it allows a better control of the polymer deposition [51]. Electrochemical polymerization is carried out with a classical three-electrode set-up in an electrochemical cell containing a monomer, a solvent and a supporting salt. The electropolymerization can be achieved either with a potentiodynamic technique such as cyclic voltammetry where the current response to a linearly cycled potential sweep between two or more set values is measured, with a potentiostatic technique where a constant potential is applied to initiate the polymerization, or with a galvanostatic technique where a constant current is applied to initiate polymerization. The potentiostatic technique allows easy control of the film thickness through Faraday's law, whereas potentiodynamic techniques lead to more homogeneous and adherent films on the electrode. Additionally, the galvanostatic technique is generally considered as the best approach since it allows to follow the growth of the conducting polymer film by monitoring the potential changes with time which reflects the conductivity.

Conducting polymers have been widely used in the area of bioanalytical and biomedical science [52,53], drug delivery [54–56], tissue engineering [57–59], and cell culture [60–62] due to their intrinsic properties and biocompatibility [63–66]. In addition, conducting polymers represent an attractive sensitive material for biosensors due to their electrical properties that allow to convert biochemical information into electrical signals. Additionally, conducting polymers can be easily modified by grafting of functional groups which offers the possibility to enhance their abilities to detect and quantify bioanalytes or to maximize the interactions between the biomolecules and the functionalized polymer. Therefore, after a short description of the electrochemical techniques used in conducting polymer-based biosensors, a series of examples of such biosensors will be described to highlight the recent advances in the field of conducting polymer-based electrochemical biosensors.

2.2. Strategies for Immobilizing Biological Sensing Elements into Conducting Polymers

Biological sensing element immobilization plays a fundamental role in the performance characteristics of biosensors since biomolecules must be directly attached to the surface of the biosensor to obtain a good sensitivity and a long operational life. The most commonly used methods to immobilize biomolecules to polymers are physical adsorption, covalent attachment and entrapment (Figure 1). The choice of immobilization strategy mainly depends on the type of biological element. Indeed, antibodies and ssDNA are preferentially immobilized by adsorption or covalent binding onto the surface of the conducting polymer films to facilitate the access of the analyte to these biorecognition molecules when entrapment is generally used to immobilize oxidoreductases within the polymer film to facilitate the electron transfer from the enzyme's redox center to the analyte solution surrounding the conducting polymer and the rapid redox reaction of electroactive species such as hydrogen peroxide generated by enzymatic catalysis.

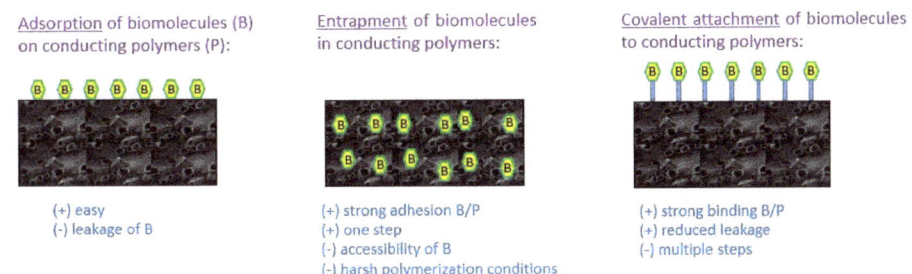

Figure 1. Strategies of immobilization of biomolecules in/on conducting polymers: advantages and drawbacks.

The method of covalent immobilization uses the functional groups of biomolecules (such as –COOH, -NH$_2$, or -SH) for binding with a conducting polymer. Thus, a biomolecule containing amino groups has the capacity to form amide bonds with a conducting polymer bearing carboxylic groups. For example, Kim et al. have developed a glucose biosensor with a conducting electrosynthesized poly(terthiophene benzoic acid) bearing benzoic acid groups which allow the immobilization of glucose oxidase (GOx) through amide bond formation [67]. Similarly, Tuncagil et al. electrosynthesized the conducting polymer 4-(2,5-di(thiophen-2-yl)-1H-pyrrol-1-yl) benzenamine to immobilize GOx through amide bonds [68]. Moreover, covalent attachment of biomolecules is frequently achieved by initial synthesis of functionalized monomers with an amino side group, followed by electrochemical polymerization of these functionalized monomers leading to conducting polymer films with interfacial attachable side groups that can be covalently bound to biomolecules containing the corresponding groups. To facilitate the formation of covalent bonds between biomolecules and polymers, crosslinking agents such as glutaraldehyde [69,70] or 1-ethyl-3-(3-dimethylaminopropyl) carbodiimide (EDC) [71,72] are commonly used. The covalent immobilization method has the benefit of providing low diffusional resistance, giving strong binding force between biomolecule and polymer, thus reducing loss of biomolecule. Therefore, these electrodes are more stable in time even if it may be difficult in some cases to retain the biomolecule activity.

The adsorption method is very simple and only consists in the physical adsorption of the biomolecule on the polymer surface. Sometimes, the presence of opposite charges into the conducting polymer and the biomolecule facilitates the immobilization of the biomolecule. Thus, negatively charged glucose oxidase was successfully adsorbed onto positively charged polyaniline-polyisoprene films at pH 4.5 to provide a material sensitive to glucose concentration changes [73]. This method has the benefit of providing small perturbation of the biomolecule native structure and function and so generally leads to very sensitive responses. However, a strong drawback is that direct physical adsorption of biomolecule on a surface generally leads to poor long-term stability of the sensor because of biomolecule leakage from the surface when changes in the environment arise (pH, ionic strength) even if the modification of the surface by a polymer film can slow this leakage [74,75].

Entrapment is another method widely used for the immobilization of enzymes [76,77], antibodies [78] or DNA [79]. It involves the preparation of an electrolyte solution containing both monomer and biomolecule, followed by the electropolymerization of the whole solution. Thus, a polymer film containing biomolecules is formed at the electrode surface. Entrapment is an interesting technique since it leads to a strong adhesion between biomolecule and polymer film in a single step. Additionally, this strategy includes the possibility of controlling the amount of entrapped biomolecules simply by controlling the thickness of the electrodeposited polymer film. Entrapment generally leads to biosensors with a good sensitivity and a long lifetime. On the contrary, entrapment can generate problems associated with inaccessibility of the embedded biomolecule. Additionally, some conducting polymers require very acidic conditions or high oxidation potential during the electropolymerization process to be prepared but these conditions are not favorable to biomolecules [80]. It is also important to note that supporting electrolytes are usually used during the electropolymerization process to increase the conductivity of the monomer solution. Besides, the electrolytes tend to compete with the biomolecules for the polymer doping sites, and so reduce the amount of biomolecule entrapped which is a problem especially for costly biomolecules. A solution to this problem is the use of biomolecules as counter-ions during the growth of the conducting polymer film to allow a more efficient entrapment as previously done with polypyrrole and GOx enzyme [81]. To enhance the incorporation of enzymes into polymers during their electropolymerization, it is also possible to use sinusoidal voltages as evidenced by Lupu et al. who developed dopamine biosensors based on tyrosinase entrapped into PEDOT film [82].

3. Electroanalytical Methods

When conducting polymers are used as sensitive material in electrochemical sensors, the capture of a target analyte to a bioreceptor immobilized in a conducting polymer generates an analytical measurable signal which is converted into an electrical signal (Figure 2). The presence of the conducting polymer is beneficial for improving sensitivity and selectivity of the biosensor while reducing the effect of interfering species. The selectivity of the biosensor strongly depends on the presence of specific interactions between the analyte and the bioreceptor when the quality of the immobilization of the bioreceptor to the conducting polymer and of the conducting polymer to the surface of the biosensor is mainly responsible for the long-term efficiency of the biosensor. The sensitivity of the biosensor depends on many factors, the main one being the intensity of the electrochemical signal generated by the reaction between the analyte, the bioreceptor and the conducting polymer. This electrochemical signal can be a change in the value of the voltage, current, conductivity/resistance, impedance, or number of electrons exchanged through an oxidation or reduction reaction leading to the fabrication of potentiometric, amperometric, conductimetric, impedimetric and voltametric biosensors, respectively.

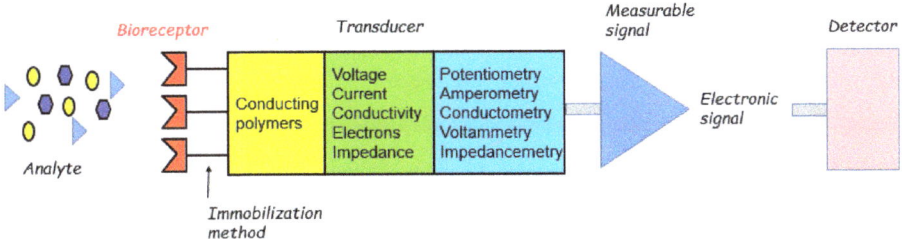

Figure 2. Detection principles of conducting polymer-based electrochemical biosensors.

3.1. Potentiometry

In potentiometric conducting polymer-based biosensors, the potential between a reference electrode and an electrode coated either with a conducting polymer and a biorecognition element is measured using a high impedance voltmeter. The conducting polymer must be sensitive to the products of a reaction involving the analyte bound to the biorecognition element. This modified electrode senses the variation in protons (or other ions) amount since potential and pH are linked by the Nernst equation, leading to a recorded analytical signal which is generally logarithmically correlated with the analyte concentration.

The most widely studied potentiometric biosensors are enzymatic biosensors that use an enzyme incorporated in the conducting polymer to catalyze a reaction producing protons. For example, in potentiometric urea enzymatic biosensors, urease is immobilized in a polymer and is used to catalyze the conversion of urea to carbon dioxide, ammonia and protons which produce a pH increase detected by the potentiometric biosensor (see Section 4.1).

3.2. Amperometry

In amperometric conducting polymer-based biosensors, the current produced during the oxidation or reduction of an electroactive biological element at a constant potential that is applied between a reference electrode and a polymer-modified electrode is measured, thus providing specific quantitative analytical information. Such biosensors are inspired by the first and simplest amperometric biosensor developed by Clark in 1956 who fabricated an amperometric oxygen sensor that produced a current proportional to the oxygen concentration when a potential of −0.6 V vs. Ag/AgCl electrode was applied to a platinum electrode [83].

The most widely studied amperometric biosensor is the glucose biosensor. In this system, glucose oxidase catalyzes the reaction of glucose with oxygen to produce gluconolactone and hydrogen peroxide. By monitoring the amount of hydrogen peroxide produced by this reaction in the presence of GOx through amperometric measurements, it is possible to determine the glucose concentration. In such biosensors, the GOx is immobilized in the conducting polymer either by electropolymerization of a solution containing a monomer and GOx or by addition of GOx in an electrodeposited conducting polymer film (see Section 4.1).

3.3. Conductometry

In conductometric conducting polymer-based biosensors, a change in electrical conductivity or resistivity is measured against the analyte concentration when a constant or sweeping potential is applied between a reference electrode and a polymer-modified electrode. To increase the sensitivity of the sensor, the conducting polymer must be highly conductive when it is charged (doped) and lowly conductive when it is neutral (dedoped), thus leading to a strong conductivity change when the conducting polymer reacts with the analyte. Furthermore, the morphology of the conducting polymer is important since the charges that are created within the backbone of the polymer must be able to interact with the surrounding environment and thus to change the polymer's conductivity. However, these biosensors often suffer from their lack of selectivity since any change in conductivity in the solution modifies their signal.

An example of conductometric biosensor was fabricated by Forzani et al. [84] who coated a pair of nanoelectrodes with PANI/GOx. Their exposure to glucose resulted in the reduction of GOx which was spontaneously reoxidized in the presence of oxygen to form hydrogen peroxide which oxidized the PANI, leading to a change in conductivity that can be monitored and used to determine the glucose concentration.

3.4. Voltammetry

In voltametric conducting polymer-based biosensors, a current is produced by sweeping the potential applied between a reference electrode and a polymer-modified electrode over a range that is associated with the redox reaction of the analyte. This redox reaction generates a change in the peak current which can be correlated with the analyte concentration, thus providing specific quantitative analytical information. All the voltametric methods that can be used, such as linear voltammetry, cyclic voltammetry, differential pulse voltammetry, or square wave voltammetry, have the advantage of providing both qualitative information deduced from the potential location of the current peak and quantitative information deduced from the peak current intensity.

For example, the detection of acetylcholine was successfully achieved using a conductive PEDOT film loaded with Fe_3O_4 nanoparticles and reduced graphene oxide since the intensity of the oxidation peak present in the cyclic voltammograms was linear to the acetylcholine concentration [85]. Similarly, the detection of serotonin in banana was done by square wave voltammetry using conducting polypyrrole/Fe_3O_4 nanocomposites [86] and the detection of danazol was performed by differential pulse voltammetry using conducting electrodeposited polyaniline [87].

3.5. Impedancemetry

Electrochemical impedance spectroscopy (EIS) is a sensitive technique for the analysis of biomolecular recognition events of specific binding proteins, nucleic acids, whole cells, antibodies or antibody-related substances, occurring at the modified surface [88–90]. In particular, many studies on impedometric biosensors are focused on immunosensors since the bonding between antibodies and antigens leads to the formation of an immunocomplex resulting in electron transfers and impedance changes (see Section 4.2). Moreover, impedometric biosensors allow direct detection of biomolecular recognition events without using enzyme labels and have the advantages of low cost, ease of use, portability and ability to perform both screening and online monitoring without being destructive. In impedometric immunosensors, conducting polymers are generally used to immobilize the antigens, commonly through covalent attachment, thus allowing the detection of the antibodies due to the high antigen-antibody affinity [91–93].

4. Conducting Polymer-Based Electrochemical Biosensors

4.1. Conducting Polymer-Based Enzyme Biosensors

Enzymatic electrochemical biosensors utilize the biospecificity of an enzymatic reaction, along with an electrode reaction that generates a current or potential for quantitative analysis. Many biomolecules such as glucose, cholesterol, or urea are important analytes due to their adverse effects on health. Enzymatic biosensors utilize the biochemical reactions between analyte and enzyme resulting in a product (hydrogen peroxide, protons, ammonium ions) that can be quantified by a transducer. In general, many oxidoreductases, for example glucose oxidase, catalyze the oxidation of substrates by electron transfer to oxygen to form hydrogen peroxide. These oxidoreductase enzymes can be immobilized on conducting polymer films and the H_2O_2 formed as a result of enzyme-catalyzed reactions induces an amperometric signal measured by the electrochemical biosensor. As a consequence, many biosensors have already been prepared that use conducting polymers as a matrix to immobilize enzymes at the surface of the biosensors.

My objective here is not to describe all the very numerous works in the field of enzymatic biosensors, but to focus on some examples of glucose sensors, illustrating the major current trends and the progress made in recent years in this research area. Indeed, in the field of biosensors, glucose biosensors have given rise to the highest number of studies due to the clinical significance of measuring blood glucose levels in patients with diabetes. Thus, some potentiometric biosensors used polyaniline films to detect pH changes due to the production of protons by oxidation of hydrogen peroxide [94,95]. However, the vast majority of classical glucose biosensors have been prepared by electropolymerization of a solution containing glucose oxidase and a monomer and used an amperometric detection. Thus, many glucose biosensors associated GOx with polyaniline or GOx with polypyrrole as extensively described in the reviews from Lai et al. [96] and Singh et al. [97], respectively. However, such conventional conducting-based glucose biosensors still present some problems such as unsatisfactory sensitivity or detection limit and relatively high applied potential. That is why more recent biosensors have substituted polymer matrix with polymer nanocomposites having the properties of increasing permselectivity, sensitivity and stability, and to decrease applied potential. Thus, conducting polymers were combined with artificial mediators such as benzoquinone derivatives [98], ferrocene derivatives [99], and Os-complex mediators [100], which were able to reoxidize the reduced GOx to its oxidized state. Thus, the released electrons from the reduced GOx are transferred to the polymer modified electrode through the redox process of the mediators. Therefore, the incorporation of a mediator leads to a better charge transport which is responsible for an enhancement of the biosensor's sensitivity. Another problem encountered in classical biosensors is that a high anodic potential (exceeding +0.6 V) is applied, leading to interference from other oxidizable substances, such as ascorbic acid, acetaminophen or uric acid. To solve this problem of interferences, it is possible to add to the sensitive layer of the biosensor a polymeric membrane (for example in Nafion or polyphenol) permeable to glucose and hydrogen peroxide but impermeable to the interfering species [101,102]. Similarly, electrocatalysts for reduction of hydrogen peroxide, such as Prussian Blue, can be used to lower reduction potential of H_2O_2 and solve the selectivity problem. Thus, Chen et al. have prepared a glucose biosensor where GOx was entrapped in a polyaniline and Prussian Blue film and which operated at the low potential of 0.0 V/SCE. This biosensor does not exhibit any interference with ascorbic acid and uric acid. It also shows a good stability, high sensitivity, rapid response, good reproducibility, and long-term stability [103].

Recently, rapid progress in the field of nanotechnology has contributed to new achievements in glucose biosensing. Indeed, association of conducting polymers with metal nanoparticles such as Au [104,105], Pt [106] or carbon materials, such as carbon nanotubes (CNT) or graphene, allowed higher GOx loading to be accessed and facilitated the electron transfer between GOX and the electrode due to their remarkable electrochemical and electrocatalytic properties. For example, Chowdhury et al. immobilized GOx onto Au nanoparticles decorated polyaniline nanowires by covalent attachment for sensing of glucose leading to lower detection limit, higher sensitivity, and greater stability than those obtained without nanoparticles [104]. Concerning carbon materials, Yuan et al. constructed a glucose biosensor based on GOx adsorbed onto a film of polyaniline containing multiwall nanotubes and Pt nanoparticles via covalent interaction with glutaraldehyde [107]. The resulting biosensor exhibited very high sensitivity because of the synergistic catalytic activity between polyaniline, multiwall nanotubes and Pt nanoparticles. Another electrochemical glucose biosensor based on glucose oxidase immobilized on a surface containing Pt nanoparticles electrodeposited on conducting poly(Azure A) previously electropolymerized on activated screen-printed carbon electrodes has been developed [108]. The resulting biosensor was validated towards glucose oxidation in real samples and further electrochemical measurement associated with the generated H_2O_2 (Figure 3). The electrochemical biosensor operated at a low potential (0.2 V vs. Ag/AgCl) and was successfully applied to glucose quantification in several real samples (commercial juices and a plant cell culture medium), exhibiting a high accuracy when compared with a classical spectrophotometric method.

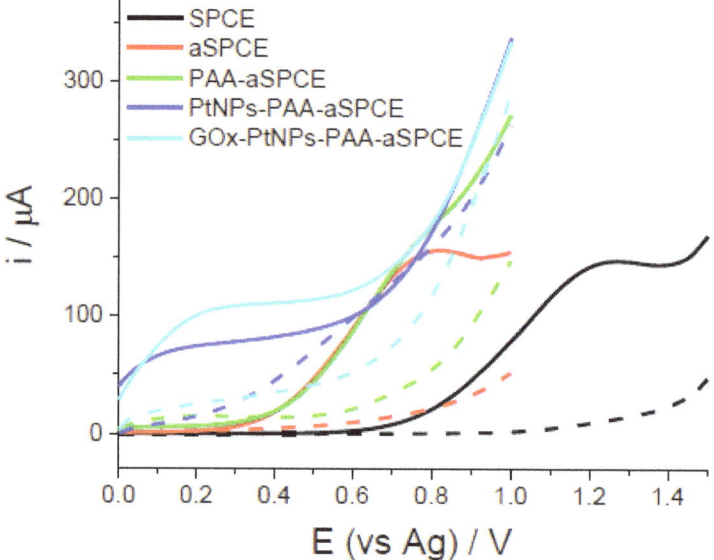

Figure 3. Anodic linear scan voltammetry responss of the different electrode modification steps in the absence (dashed lines) and the presence of 5M of H_2O_2 (solid lines) in 0.1 M phosphate buffer solution. SPCE: Screen-printed carbon electrodes, aSPCE: activated SPCE, PAA: Poly(Azur A), PtNPs: Platinum Nanoparticles, GOx: glucose oxidase. Reproduced with permission of [108], Copyright 2020, MDPI.

Enzymatic electrochemical biosensors can also be based on a conducting polymer and metal oxide nanoparticles. For example, a biosensor based on lipase was developed for amlodipine besylate (AMD) drug using a mixture of polyaniline, iron oxide and gelatin (Figure 4). After preparation of the sensitive material (step 1), the enzyme was entrapped in the biocomposite matrix film with the aid of a glutaraldehyde cross-linking reagent (step 2) to establish the immobilization of the lipase (step 3) [109]. Cyclic voltammetry (A) and impedometry (B) were then used for detection experiments which proved that such material was a good candidate for the construction of a sensitive biosensor for AMD analysis (Figure 4b). Similarly, another glucose biosensor was prepared from polyaniline, nickel oxide nanoparticles and graphene oxide in order to exploit the synergy between those kinds of materials The biosensor showed a good sensitivity, a low detection limit of 0.5 mM, and a good selectively since glucose could be detected in the presence of common interfering species such as ascorbic acid, uric acid and dopamine. [110].

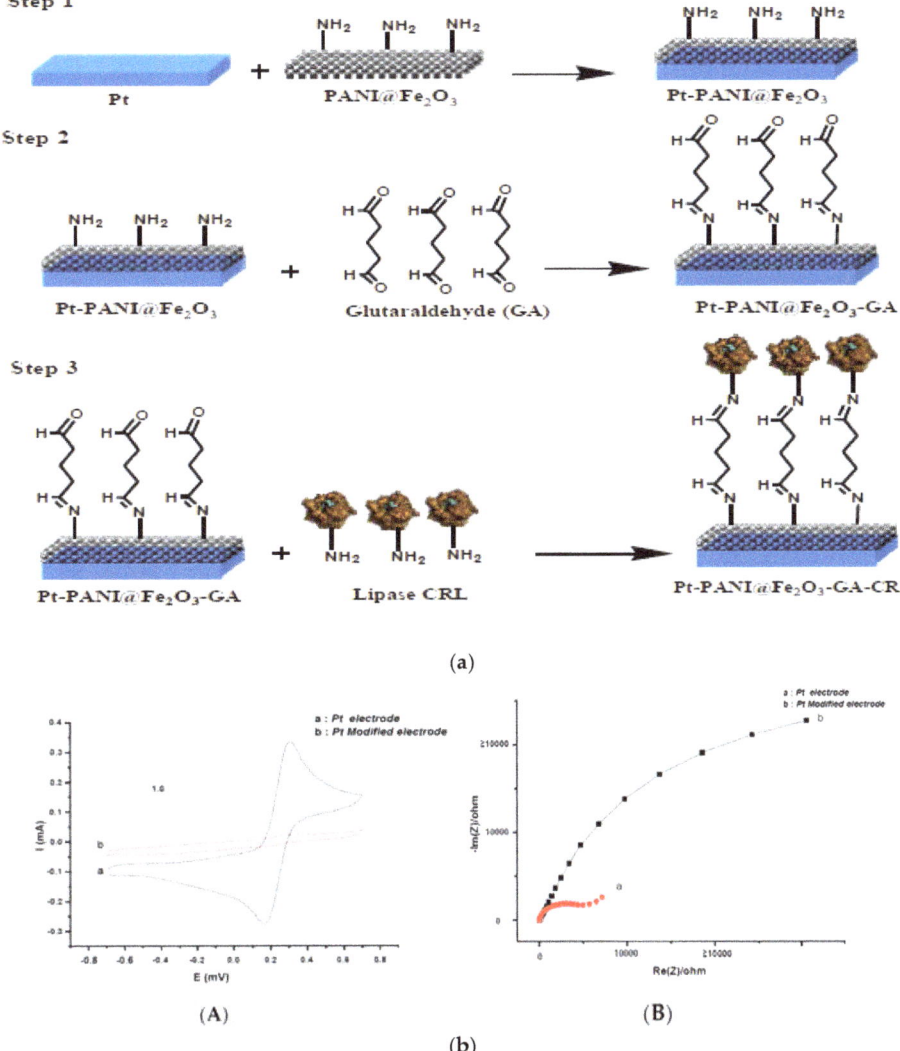

Figure 4. (**a**) Schematic representation of the biosensor elaboration using PANI@Fe$_2$O$_3$; (**b**) Voltammograms at 50 mV/s (**A**) and Diagram of Nyquist at −200 mV potential of Pt bare electrode and Pt electrode modified with an enzymatic membrane in 0.1 M phosphate buffered saline solution in the presence of Fe(CN$_6$)$^{3-/4-}$. (**B**) impedometry Reproduced with permission of [109], Copyright 2018, MDPI.

At nanoscale, conducting polymer are processable and it is possible to prepare polymer nanostructures, using chemical template-based syntheses or template free methods, that allow higher GOx loading and more sensitive response to glucose. For example, a glucose biosensor has been reported which is based on GOx electrochemically entrapped into the inner wall of highly ordered polyaniline nanotubes synthesized using anodic aluminium oxide as template [111]. This biosensor-enhanced electrocatalysis as indicated by its high sensitivity (97 µA·mM^{-1}·cm^{-2}), fast response time (3 s), efficient preservation of enzyme activity, and effective discrimination to common interfering species. The strategy consisting in the use of polymer nanowires was also chosen by Xu et al. who developed a glucose biosensor, based on the modification of well-aligned polypyrrole nanowires array with Pt nanoparticles and subsequent surface adsorption of GOx [112]. This biosensor showed evidence of direct electron transfer due mainly to modification incorporating Pt nanoparticles and allowed either potentiometric or amperometric detection. Using another strategy, Komathi et al. prepared polyaniline nanoflowers with protruded whiskers at the edge of the flowers using cetyltrimethylammonium bromide as a soft template and fine tuning the graft co-polymerization conditions. These nanostructures exhibited wider linear concentration range, low detection limit, and high sensitivity compared to most of the previously reported classical enzyme glucose sensors [113].

An ultimate goal of the biosensors is to eliminate the usage of the mediator to lower fabrication cost and complexity while increasing the durability of the biosensor. Therefore, the third-generation biosensors based on the direct electron transfer from immobilized enzyme to the working electrode are a more progressive type of sensor. Such direct electron transfer have been evidenced from redox enzymes to electrode in conducting polymer-based biosensors by Ramanivicius et al. who reported for the first time that direct electron-transfer processes between a polypyrrole entrapped quinohemoprotein alcohol dehydrogenase from Gluconobacter sp. 33 and a platinum electrode take place via the conducting-polymer network [114]. The cooperative action of the enzyme-integrated prosthetic groups is assumed to allow this electron-transfer pathway from the enzyme's active site to the conducting-polymer backbone. This electron-transfer pathway leads to a significantly increased linear detection range of an ethanol sensor. Since this work, dehydrogenase based bioelectrocatalysis has been increasingly exploited in order to develop electrochemical biosensors with improved performances since dehydrogeases are able to directly exchange electrons with an appropriately designed electrode surface, without the need for an added redox mediator, allowing bioelectrocatalysis based on a direct electron transfer process [115]. A direct electron transfer can also occur from immobilized glucose oxidase via grafted and electropolymerized 1,10-phenanthroline [116]. Such polymer-modified biosensor showed superior electron transfer to/from flavine adenine dinucleotide cofactor of GOx as well as an excellent selectivity towards glucose and a good operational-stability. Similarly, a biosensor based on electrodeposited polycarbazole was fabricated and exhibited good electrocatalytic activity toward enzymatic glucose sensors with a high sensitivity, a wide linear range of detection up to 5 mM due to direct electron transfer from the enzyme to electrode and direct glucose oxidation on the electrode [117]. Table 2 summarizes the performances of these conducting polymer-based glucose amperometric biosensors.

Table 2. Comparison of conducting-polymer-based glucose amperometric biosensor performances.

Active Layer	Linear Range	Sensitivity	Detection Limit	Stability	Real Samples?	Ref.
Polypyrrole-CNT-chitosan	1–4.7 mM	2860 µA mM^{-1} cm^{-2}	5.0 µM	45 days	serum	[101]
Polypyrrole-CNT	1–4.1 mM	54.2 µA mM^{-1} cm^{-2}	5.0 µM	45 days	serum	[102]
Polyaniline-Prussian Blue	2–1.6 MM	99.4 µA mM^{-1} cm^{-2}	0.4 µM.	15 days	serum	[103]
Polyaniline-Au NP	1–20 mM	14.6 µA mM^{-1} cm^{-2}	1.0 µM	—	—	[104]
Polyaniline-Pt NP	0.01–8 mM	96.1 µA mM^{-1} cm^{-2}	0.7 µM	—	—	[106]
Polyaniline-CNT	3–8.2 mM	16.1 µA mM^{-1} cm^{-2}	1.0 µM	48 days	serum	[107]
Poly(Azure A)-Pt NP	0.02–2.3 mM	42.7 µA mM^{-1} cm^{-2}	7.6 µM	3 month	fruit juice	[108]
Polyaniline-Graphene-NiO$_2$	0.02–5.56 mM	376.2 µA mM^{-1} cm^{-2}	0.5 µM	—	serum	[110]
Polyaniline	0.01–5.5 mM	97.2 µA mM^{-1} cm^{-2}	0.3 µM	15 days	urine	[111]
Polypyrrole-Pt NP	0.1–9 mM	34.7 µA mM^{-1} cm^{-2}	27.7 µM	—	—	[112]
Polyaniline-nanodiamonds	1–30 mM	2.03 mA mM^{-1} cm^{-2}	18.0 µM	30 days	serum	[113]
Polycarbazole	0.01–5 mM	14.0 µA mM^{-1} cm^{-2}	0.2 µM	—	—	[117]

4.2. Conducting Polymer-Based Immunosensors

An immunosensor is a type of affinity solid-state based biosensor in which a specific target analyte, antigen (Ag), is detected by formation of a stable immunocomplex between antigen and antibody as a capture agent (Ab) due to the generation of a measurable signal. Thanks to the strong affinity between antigen and antibody, electrochemical immunosensors are among the most promising bioanalytical sensors. Another advantage of the antibody-based recognition method is that the target analyte, the antigen, does not need to be purified prior to detection contrary to enzymes for example. Thus, a variety of conducting polymer-based electrochemical biosensors have been developed in recent years that showed promising sensing performances.

For example, Grennan et al. fabricated an amperometric immunosensor allowing a very low level detection of atrazine (0.1 ppb) using recombinant single-chain antibody fragments electrostatically attached to classical polyaniline associated with poly(vinylsulphonic acid) which enables direct mediatorless coupling to take place between the redox centers of antigen-labelled horseradish peroxidase and the electrode surface [118]. Similarly, Grant et al. reported the fabrication of an impedimetric immunosensor based on the direct incorporation of antibodies (anti-BSA) into polypyrrole films that allows the detection of BSA proteins with a linear response from 0 to 75 ppm [91]. Darain et al. synthesized a more original terthiophene monomer having a carboxylic acid group, 5,2′:5′2″-terthiophene-3′-carboxylic acid, and used it to immobilize the antibody monoclonal anti-vitellogenin (Vtg) through covalent amine bonds. The resulting layer allowed the detection of vitellogenin, a biomarker for xenobiotic estrogens responsible for causing endocrine disruption through antibody-antigen interactions with high selectivity and sensitive response to Vtg [119]. Similarly, Aydin et al. synthesized poly(2-thiophen-3-yl-malonic acid), an original polythiophene derivative bearing two acid side groups per monomer allowing the immobilization of anti-Interleukin-1β antibody through amide bonds after EDC-NHS treatment [92]. This sensitive layer was then used to detect Interleukin 1β in human serum and saliva by impedometric detection leading to low detection limit (3 fg/mL), good specificity, reproducibility, and stability. The immunosensor was applicable for detection of IL-1β samples.

Recently, Wang et al. proposed an immunosensor for the detection of the tumor marker neuron specific enolase (NSE) based on a complex and original sensitive layer [120]. Indeed, they prepared a hydrogel containing polypyrrole and polythionine along with GOx as a doping agent, and gold nanoparticles used to enhance the conductivity and provide a binding surface for the antibody, anti-neuron specific enolase (anti-NSE). Moreover, glucose was added to the analyte solution to react with GOx, and so generate H_2O_2 which amplified the biosensor's response. Square wave voltammetry was then used to detect NSE levels, leading to a low detection limit (0.65 pg/mL) and a wide linear range. Another group developed an immunosensor for the detection of carcinoma antigen-125 (CA 125) which was based on a hydrogel composed of polypyrrole, polythionine, gold nanoparticles, and phytic acid used as a polymer crosslinker to increase hydrophilicity and provide an antifouling capability. The as-prepared immunosensor exhibited a wide linear range from 0.1 mU/mL to 1 kU/mL, a low limit of detection (1.25 mU/mL), a high sensitivity and an excellent specificity [121]. Another immunosensor dedicated to the detection of CA 125 and based on electrodeposited poly(anthranilic acid) and gold nanoparticles was prepared by Taleat et al. [94]. The monoclonal anti-CA 125 antibodies were covalently immobilized on poly(anthranilic acid) using EDC-NHS and labeled with gold nanoparticles before being used to capture and detect CA 125 using electrochemical impedimetric measurements with a good sensitivity and reproducibility which matches the request of clinical needs (cancer antigens CA 125 are cancer biomarkers). Similarly, Shaikh et al. developed an impedimetric immunosensor for the sensitive, specific, and label-free detection of human serum albumin (HSA, a valuable clinical biomarker for the early detection of chronic kidney disease) in urine. To enable efficient antibody immobilization and improved sensitivity, the carbon working electrode was sequentially modified with electropolymerized polyaniline and electrodeposited gold nanocrystals (Figure 5). Indeed, polyaniline (b) and Au nanocrystals (c) were successively electrodeposited on the screen-printed working electrode (a). Then, oxidized HSA antibody (d) and BSA (e) were immobilized onto the polyaniline-Au layer. Finally, the evolution of impedance (g) was used to quantify the amount of HSA (f). The normalized impedance variation during immunosensing increased linearly with HSA concentration and the biosensor displayed highly repeatable and highly specific response to HSA concentration [122].

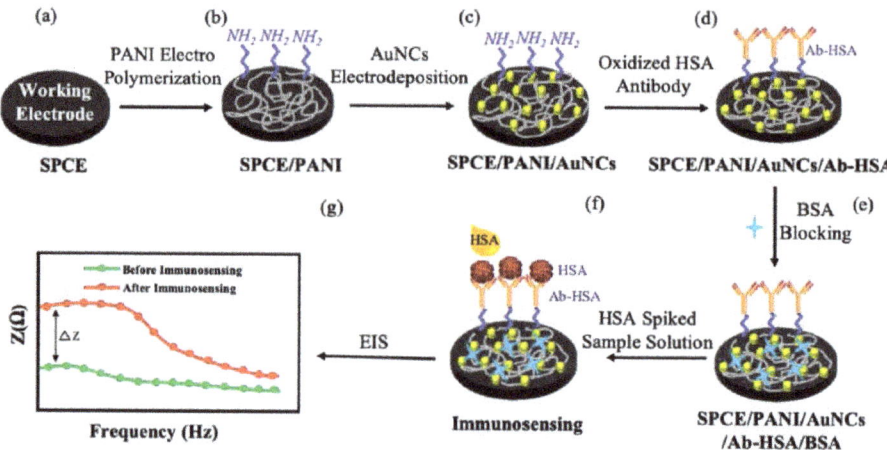

Figure 5. Schematic representation of the protocol for surface modification and immunosensing. SPCE: Screen-printed carbon electrodes, PANI: polyailine, AuNCs: gold nanocrystals, HSA: human serum albumin, Ab-HSA: anti-HAS, EIS: electrochemical impedance spectroscopy. (**a**) screen-printed working electrode ; (**b**) polyaniline ; (**c**) Au nanocrystals (**d**) oxidized HSA antibody (**e**) BSA (**g**) the evolution of impedance (**g**) HSA Reproduced with permission of [122], Copyright 2019, MDPI.

Recently, the first immunosensors based on conducting polymer nanostructures using template methods have been developed. For example, an immunosensor for the determination of alpha-fetoprotein (AFP) was fabricated based on the three-dimensional macroporous polyaniline doped with poly (sodium 4-styrene sulfonate) by using a hard-template method [123]. The 3D macroporous PANI possessed large surface area, high conductivity and many functional groups, which allowed the immobilization of anti-AFP. Based on differential pulse voltammetry measurements, the prepared AFP immunosensor showed a wide linear range for AFP from 0.01 to 1000 pg/mL, with a detection limit of 3.7 fg/mL. Table 3 summarizes the performances of these conducting polymer-based immunosensors.

Table 3. Comparison of conducting polymer-based immunosensor performances.

Active Layer	Target	Detection Mode	Linear Range	Detection Limit	Ref.
Polyaniline-poly(vinylsulfonic acid)	atrazine	amperometry	0.12–5 µM	0.1 µg/L	[118]
Polythiophene derivative (with—COOH groups)	Carp vittelogenin	impedometry	1–8 µg/L	0.42 µg/L	[119]
Polypyrrole-polythionine	neuron-specific enolase	voltammetry	0.001–100 pg/mL	0.65 pg/mL	[120]
Polypyrrole-polythionine	carcinoma antigen-125	voltammetry	0.0001–1000 U/mL	0.00125 U/mL	[121]
Polyaniline/Au nanocrystals	human serumalbumin	voltammetry + impedometry	3–300 µg/mL	3 µg/mL	[122]
Polyaniline-poly(sodium styrene sulfonate)		voltammetry	0.01–1000 pg/L	3.7 fg/mL	[123]

4.3. Conducting Polymer-Based DNA Biosensors

There is a great interest in the development of easy-to-use DNA biosensors, since detection of specific DNA sequences is of great importance in medical research and clinical diagnosis, in particular for DNA diagnostics and gene analysis. DNA biosensors generally rely on the immobilization of a single stranded DNA (ssDNA) probe onto a surface to recognize its complementary DNA target sequence by hybridization. Conducting polymers are frequently used for fabrication of DNA biosensors since the immobilization of oligonucleotide (ODN) probes onto conducting polymers generally provides an electrochemical response which allows a direct way to detect hybridization events.

Adsorption method for probe DNA immobilization into conducting polymers offers the simplest methodology but suffers from poor stability and response if not properly optimized. One of the most convincing studies using the adsorption method has been carried out by Dutta et al. who used few layered MoS_2 nanosheets blended with conducting polyaniline to perform DNA sensing via differential pulse voltammetry technique [124]. This biosensor worked well even at concentrations as low as 10^{-15} M of target DNA and showed highly satisfactory results in case of serum samples. It is also possible to use ODN entrapment in conducting polymer films even if this method is not the most widely used in the area of DNA sensors since it can be difficult for the DNA target to access the entrapped ODN. However, some works exist such as the one of Eguiluz et al. who entrapped an ODN probe in a polypyrrole film during its electropolymerization, leading to a low detection limit of *Alicyclobacillus acidoterrestris* DNA [125]. Another strategy was used by Tlili et al. who synthesized a polypyrrole derivative to facilitate the covalent attachment of an ODN. Indeed, the copolymer poly [3-acetic acid pyrrole, 3-*N*-hydroxyphthalimide pyrrole)] was electropolymerized, then a direct chemical substitution of the leaving *N*-hydroxyphthalimide group by the oligonucleotide was realized leading to the formation of amide bonds between the ODN probe bearing a terminal amino group on its 5' phosphorylated position and the copolymer film [126]. The hybridization reactions with the DNA complementary target and non-complementary target were then investigated by both amperometric and impedimetric analyses which demonstrated a good sensitivity and low detection limit (1 pmol).

The elaboration of nanocomposites by combination of conducting polymers and metallic or carbon materials can also be used to covalently attach ODN and enhance the response of DNA biosensors. Thus, Wilson et al. fabricated a DNA biosensor in which a DNA labelled at 5' end using 6-mercapto-1-hexhane was covalently immobilized by the Au-thiol chemistry onto a PPy-PANI-Au film obtained by successive chemical oxidation of PPy and PANI, followed by electrodeposition of gold [127]. This association of conducting polymers with Au nanoparticles increased the conductivity and provided an enhancement of the hybridization efficiency. Similarly, the electrochemical DNA hybridization sensing of bipolymer polypyrrole and PEDOT functionalized with Ag nanoparticles has been investigated [128]. DNA labeled at 5' end using 6-mercapto-1-hexhane was immobilized on the PPy-PEDOT-Ag surface, and the resulting impedometric biosensor effectively allowed the detection of target DNA sequences with a wide dynamic detection range and a low detection limit of 5.4×10^{-15} M. It is also possible to combine polymers and carbon nanotubes as done by Xu et al. who prepared an impedometric DNA biosensor by using a composite material of polypyrrole electropolymerized in the presence of carboxylic groups ended multiwalled carbon nanotubes [129]. Amino group ended single-stranded DNA probe was linked onto the PPy/MWNTs-COOH using carbodiimide for crosslinking amine and carboxylic acid group. The PPy/MWNTs-COOH film exhibited a good electronic transfer property and a large specific surface area and led to a high sensitivity and selectivity of this biosensor. In this work, PPy did not consist in a film deposited on a substrate but it consisted of nanotubes. This is a recent trend to use nanostructured polymers for biosensing applications and PANI nanotubes have also be used in DNA biosensors [130] as well as PANI nanowires [131] or PANI nanofibers [132,133]. Chang et al. prepared conducting polyaniline nanotubes to induce a signal enhancement compared to classical polyaniline [130]. A PANI nanotube array with a highly organized structure was fabricated under a well-controlled nanoscale dimension on a graphite electrode using a nanoporous layer as a template, and 21-mer oligonucleotide probes were immobilized on these nanotubes. The electrochemical results showed that the DNA biosensor detected the target oligonucleotide at a concentration as low as 1.0 fM. In addition, this biosensor demonstrated good capability of differentiating the perfect matched target ODN from one-nucleotide mismatched ODN even at a low concentration. Another electrochemical DNA biosensor based on electrochemically fabricated polyaniline nanowires and methylene blue was used for DNA hybridization [131]. Nanowires of conducting polymers, with diameters in the range from 80 to 100 nm, were directly synthesized through an electrochemical deposition procedure. Oligonucleotides with phosphate groups at the 5′ end were covalently linked onto the amino groups of polyaniline nanowires on the electrode. The hybridization events were monitored with differential pulse voltammetry measurement using methylene blue as an indicator. The approach described here can effectively discriminate complementary from non-complementary DNA sequence, with a detection limit of 1.0×10^{-12} mol/L of complementary target, suggesting that the polyaniline nanowires hold great promises for sensitive electrochemical biosensor applications. Du et al. have electrodeposited reduced graphene oxide on polyaniline nanofibers, and the formed nanocomposites were applied to bind ssDNA probe via the non-covalent assembly [132]. After the hybridization of ssDNA probe with complementary DNA, the response of the biosensor changed obviously, and allowed selective detection of the sequence-specific DNA of cauliflower mosaic virus gene with a detection limit of 3.2×10^{-14} mol/L. Finally, polyaniline and graphene composite nanofibers (ranging from 90 to 360 nm in diameter) were prepared by oxidative polymerization in the presence of a solution containing poly(methyl vinyl ether-alt-maleic acid) (Figure 6). The composite nanofibers with an immobilized DNA probe were used for the detection of *Mycobacterium tuberculosis* by using differential pulse voltammetry method leading to a detection range of 10^{-6}–10^{-9} M with the detection limit of 7.8×10^{-7} M under optimum conditions [133]. These results show that the composite nanofibers have a great potential in a range of applications for DNA sensors. Table 4 summarizes the performances of these conducting polymer-based DNA biosensors.

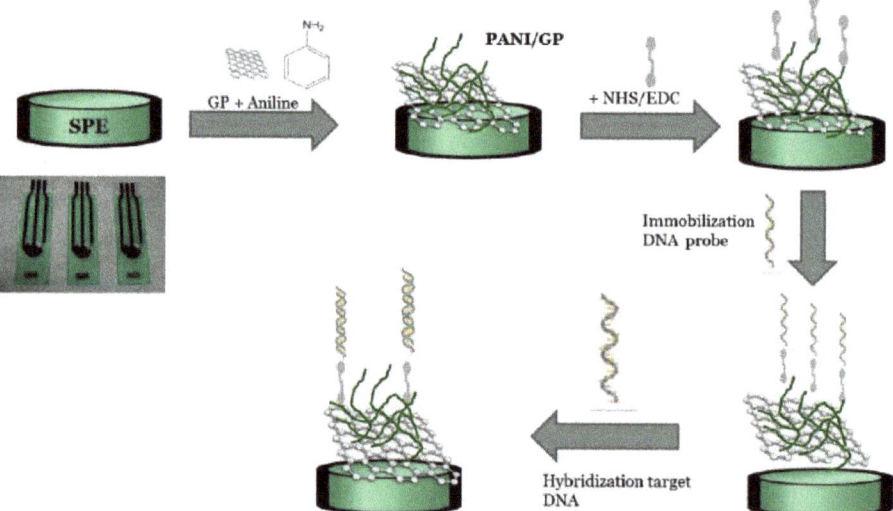

Figure 6. Schematic illustration of the stepwise electrochemical fabrication process for DNA biosensor. Reproduced with permission of [133], Copyright 2017, MDPI.

Table 4. Comparison of conducting polymer-based DNA biosensor performances.

Active Layer	Detection Mode	Linear Range	Detection Limit	Ref.
Polyaniline–MoS$_2$	voltammetry	10^{-15}–10^{-6} M	10^{-15} M	[124]
Polypyrrole–Au and Ag NPs	voltammetry	7–150 nM	7 nM	[125]
poly [3–acetic acid pyrrole,3–N–hydroxyphthalimide pyrrole)]	impedometry	0.05–5.5 nM	1 pM	[126]
Polypyrrole–Polyaniline–	impedometry	10^{-13}–10^{-6} M	10^{-13} M	[127]
Polypyrrole–PEDOT–Ag NP	impedometry	10^{-15}–10^{-11} M	5×10^{-15} M	[128]
Polypyrrole–CNT–COOH	impedometry	10^{-12}–10^{-7} M	5×10^{-12} M	[129]
Polyaniline	voltammetry	10^{-15}–10^{-12} M	10^{-15} M	[130]
Polyaniline–methylene blue	voltammetry	10^{-12}–10^{-10} M	10^{-12} M	[131]
Polyaniline–graphene	voltammetry	10^{-13}–10^{-7} M	3×10^{-14} M	[132]
Polyaniline–graphene	voltammetry	10^{-9}–10^{-6} M	8×10^{-7} M	[133]

4.4. Conducting Polymer-Based Whole Cell Biosensors

Whole cells are more complex biological recognition elements than isolated components such as enzymes, but they offer many advantages including low cost (no cost for isolation process), less time consuming due to reduced processing, better resistance to pH and temperature. Therefore, whole cells hold the promise of allowing significant progress in the field of cell-based electrochemical biosensors having a wide range of applications in pharmacology, medicine, cell biology, toxicology, and neuroscience [134].

Until now, conducting polymers have been mainly used as conductive scaffolds to enhance the adhesion and proliferation of cells on substrates [135–138]. Thus, El-Said et al. electrodeposited a film of conductive polyaniline on the ITO electrode of a cell-based chip which was used to measure the cellular electrochemical properties of HeLa carcinoma cells and monitor the effects of different anticancer drugs on the cell viability [137]. To go further and to develop interface electrical devices with neural cells allowing long-term implantation, some research groups develop nanoelectrode arrays incorporating nanostructured conducting polymers. For example, Nguyen-Vu et al. achieved the culture of neural cells on electrodeposited vertically aligned polypyrrole nanoarrays that can serve as a 3D interface between neural tissues and electronic biosensors [139]. In another study, polypyrrole nanowires electrodeposited in highly ordered nanoporous alumina substrates were used to immobilize cancer cells. These polypyrrole nanowires were found to exhibit better cell adhesion and proliferation than traditional culture substrates showing the potential of biocompatible electroactive polymer for both healthy and cancer cell cultures applications [140].

There are also a few examples of biosensors using both whole cells (mainly microbial cells) and conductive polymers. Thus, a rapid and sensitive determination of glucose in biological samples was performed using conducting polypyrrole and whole *Aspergillus niger* microbial cells, rather than pure enzymes, as bioreceptors. The use of whole microbial cells enabled a reduction in the cost of the biosensor and an improvement of the adaptability of the biosensor to adverse conditions [141]. Another amperometric biosensor based on *Gluconobacter oxydans* whole cells and electrodeposited poly(10-(4H-dithiyeno [3,2-b:2',3'-d]pyroll-4-il)decan-1-amine) was fabricated for the detection of glucose and exhibited good analytical performances in terms of sensitivity and dynamic range [142]. In another study, the same *Gluconobacter oxydans* whole cells were immobilized on an electrodeposited poly(4,7-di(2,3)-dihydrothienol[3,4-b][1,4]dioxin-5-yl-benzo[1,2,5]thiadiazole) film used to sense glucose since the respiratory activity of the cells was found to be directly proportional to the glucose concentration [143]. Similar results were obtained with another glucose biosensor designed by the same researchers and associating *Gluconobacter oxydans* whole cells with electrodeposited poly(4-amino-N-(2,5-di(thiophen-2-yl)-1H-pyrrol-1-yl)benzamide) [144,145]. Furthermore, an efficient conductometric urea biosensor was fabricated which used the change in resistivity generated by an increase of the pH due to the catalytic action of urease contained in the whole *Brevibacterium ammoniagenes* cells previously immobilized in a polystyrene sulphonate–polyaniline (PSS–PANI) conducting film [146].

4.5. Biosensors Based on Molecularly Imprinted Polymers

A new trend in the area of biosensors concerns the use of molecularly imprinted polymers (MIPs). MIPs are biomimetic receptors that are synthetically prepared by polymerizing monomers in the presence of the target analyte (used as a template). Upon template removal, this process generated a three-dimensional polymer matrix that provides cavities (biomimetic receptors) with the correct size, shape, and electrostatic environment to specifically interact with the molecular target.

Thus, the group of Ramanavicius electrodeposited a polypyrrole layer molecularly imprinted by caffeine and studied its properties [147]. Using quartz crystal microbalance, they demonstrated that the equilibrium of the interaction between the MIP and dissolved caffeine was shifted towards the formation of MIP/caffeine complex while the equilibrium for the interaction of MIP and theophylline was shifted towards dissociation of MIP/theophylline complex. Therefore, the obtained MIP evidenced much higher selectivity towards caffeine in comparison with the selectivity towards its homologue-theophylline. Additionally, an imprinted amperometric biosensor based on polypyrrole-sulfonated graphene/hyaluronic acid-multiwalled carbon nanotubes was fabricated for sensitive detection of tryptamine [148]. The biosensor was based on MIPs previously synthesized by electropolymerization using tryptamine as the template, and para-aminobenzoic acid as the monomer. The presence of the MIP induced an enhancement of the current response of the biosensor. The good selectivity of the sensor allowed discrimination of tryptamine from interferents (tyramine, dopamine and tryptophan). Another electrochemical sensor was developed for the recognition and detection of epinephrine by combining a MIP, silica nanoparticles and multiwalled carbon nanotubes [149]. A molecular imprinted polypyrrole film was electropolymerized on the surface of a glassy carbon electrode modified with silica nanoparticles and carbon nanotubes in the presence of epinephrine. With the etching of silica nanoparticles, the obtained amperometric biosensor exhibited a multiporous network structure which increased the efficiency of imprinted sites of the biosensor. The resulting MIP-based biosensor showed high sensitivity, good selectivity and reproducibility for epinephrine determination. Similarly, clopidol-imprinted polypyrrole films were electrochemically prepared on screen printed carbon electrodes in aqueous solutions of pyrrole and clopidol [150]. The clopidol template molecules were successfully trapped in the polypyrrole film where they created artificial recognition sites. After extraction of the template, the polypyrrole film acted as a MIP for the specific and selective recognition of clopidol. Using differential pulse voltammetry, the calibration curve of the biosensor was found to be linear for a wide concentration range, sensitive, stable, and reproducible without any influence of interferents existing in real samples. MIPs were also used to fabricate immunosensors. For example, an AFP immunosensor based on polythionine and gold nanoparticles coated by a polydopamine-AFP MIP was fabricated [151]. Indeed, a polydopamine–AFP complex was electropolymerized on a polythionine/Au nanoparticles film, applying AFP as template and dopamine as imprinted monomers. After elution, the specific cavities served to adsorb the target molecules. Using differential pulse voltammetry detection, the peak current decreased with the increase in concentration of AFP, and the linear response range of the biosensor was from 0.001 ng/mL to 800 ng/mL with a low detection limit of 0.8 pg/mL. Another immunosensor based on a MIP was developed to detect simultaneously prostate-specific antigen (PSA) and myoglobin (Myo) in human serum and urine samples. Thus, target proteins were attached covalently to 3,3′-dithiodipropionic acid di(*N*-hydroxysuccinimide ester) previously deposited on a gold substrate. The MIP was then fabricated on this surface using acrylamide as monomer, *N*,*N*′-methylenebisacrylamide as a crosslinker, and PSA and Myo as the templates, respectively [152]. After that, a nanocomposite was synthesized based on the decorated magnetite nanoparticles with multi-walled carbon nanotubes, graphene oxide and specific antibody for PSA. The ability of proposed biosensor to detect PSA and Myo simultaneously with high sensitivity and specificity offers an opportunity for a new generation of immunosensors.

5. Conclusions

This review presents an overview of the diverse strategies used for developing electrochemical biosensors based on conducting polymers and outlines the significant advances in this field. Indeed, conducting polymers have many advantages including their charge transport properties and their chemical versatility that can be used to fabricate efficient biosensors through potentiometric, amperometric, conductometric, voltametric and impedometric detection. Additionally, conductive polymers with functional groups can be synthesized and used to facilitate the immobilization of biorecognition molecules through covalent attachment which is the most commonly used method to immobilize biomolecules, but adsorption or entrapment are also often used. As a result, conducting polymers are now considered as good sensitive materials for the development of selective, specific, and stable sensing devices. However, electrodeposited polymers still have many unexplored possibilities, and so a lot of future research will probably be dedicated to the development of new polymer-based biosensors. Another promising way for the future is the nanostructuration of electrodeposited polymers since the electrosynthesis of polymer nanowires or nanotubes recently led to strong improvements in the sensing properties of conducting polymers. In addition, the landscape for hybrid conducting polymer systems combining polymers and conducting inorganic materials, especially metallic nanoparticles and carbon nanomaterials, is rich in potential with numerous promising materials, each with their own chemical, electrical and physical properties, yet to be explored for biosensing.

Funding: This research received no external funding.

Conflicts of Interest: The authors declare no conflict of interest.

References

1. Shirakawa, H.; Louis, E.J.; MacDiarmid, A.G.; Chiang, C.K.; Heeger, A.J. Synthesis of electrically conducting organic polymers: Halogen derivatives of polyacetylene, $(CH)_x$. *J. Chem. Soc. Chem. Commun.* **1977**, 578–580. [CrossRef]
2. Diaz, A. Electrochemical preparation and characterisation of conducting polymers. *Chem. Scr.* **1981**, *17*, 145–148.
3. Diaz, A.F.; Kanazawa, K.K. Electrochemical polymerisation of pyrrole. *J. Chem. Soc. Chem. Commun.* **1979**, 635. [CrossRef]
4. Kanazawa, K.K.; Diaz, A.F.; Geiss, R.H.; Gill, W.D.; Kwak, J.F.; Logan, J.A. 'Organic metals': Polypyrrole a stable synthetic 'metallic' polymer. *J. Chem. Soc. Chem. Commun.* **1979**, 854. [CrossRef]
5. Le, T.H.; Kim, Y.; Yoon, H. Electrical and Electrochemical Properties of Conducting Polymers. *Polymers* **2017**, *9*, 150. [CrossRef]
6. Tsukamoto, J. Recent advances in highly conductive polyacetylene. *Adv. Phys.* **1992**, *41*, 509–546. [CrossRef]
7. Tsukamoto, J.; Takahashi, A.; Kawasaki, K. Structure and electrical properties of polyacetylene yielding a conductivity of 10^5 S/cm. *Jpn. J. Appl. Phys.* **1990**, *29*, 125. [CrossRef]
8. Patois, T.; Lakard, B.; Martin, N.; Fievet, P. Effect of various parameters on the conductivity of free standing electrosynthesized polypyrrole films. *Synth. Met.* **2010**, *160*, 2180–2185. [CrossRef]
9. Zhang, Y.; de Boer, B.; Blom, P.W.M. Controllable molecular doping and charge transport in solution-processed polymer semiconducting layers. *Adv. Funct. Mater.* **2009**, *19*, 1901–1905. [CrossRef]
10. Guimard, N.K.; Gomez, N.; Schmidt, C.E. Conducting polymers in biomedical engineering. *Prog. Polym. Sci.* **2007**, *32*, 876–921. [CrossRef]
11. Ahlskog, M.; Reghu, M.; Heeger, A.J. The temperature dependence of the conductivity in the critical regime of the metal-insulator transition in conducting polymers. *J. Phys. Condens. Matter* **1997**, *9*, 4145–4156. [CrossRef]
12. Aleshin, A.; Kiebooms, R.; Menon, R.; Wudl, F.; Heeger, A.J. Metallic conductivity at low temperatures in poly (3,4-ethylenedioxythiophene) doped with PF_6. *Phys. Rev. B* **1997**, *56*, 3659–3663. [CrossRef]

13. Lee, K.; Cho, S.; Heum Park, S.; Heeger, A.J.; Lee, C.W.; Lee, S.H. Metallic transport in polyaniline. *Nature* **2006**, *441*, 65–68. [CrossRef] [PubMed]
14. Wataru, T.; Shyam, S.P.; Masaki, F.; Keiichi, K. Cyclic step-voltammetric analysis of cation-driven and anion-driven actuation in polypyrrole films. *Jpn. J. Appl. Phys.* **2002**, *41*, 7532–7536.
15. Paul, E.W.; Ricco, A.J.; Wrighton, M.S. Resistance of polyaniline films as a function of electrochemical potential and the fabrication of polyaniline-based microelectronic devices. *J. Phys. Chem.* **1985**, *89*, 1441–1447. [CrossRef]
16. Saxena, V.; Malhotra, B.D.; Menon, R. Charge transport and electrical properties of doped conjugated polymers. In *Handbook of Polymers in Electronics*; Malhotra, B.D., Ed.; Rapra Technology Limited: Shrewsbury, Shropshire, UK, 2002; pp. 3–65.
17. Wan, M. *Conducting Polymers with Micro or Nanometer Structure*; Springer: New York, NY, USA, 2008; pp. 1–13.
18. Li, Y. Conducting polymer. In *Organic Optoelectronic Materials*; Li, Y., Ed.; Springer International Publishing: New York, NY, USA, 2015; pp. 23–50.
19. Bredas, J.L.; Chance, R.R.; Silbey, R. Theoretical Studies of Charged Defect States in Doped Polyacetylene and Polyparaphenylene. *Mol. Cryst. Liq. Cryst.* **1981**, *77*, 319–332. [CrossRef]
20. Scrosati, B. Electrochemical Properties of Conducting Polymers. *Prog. Solid State Chem.* **1988**, *18*, 1–77. [CrossRef]
21. Beaujuge, P.M.; Reynolds, J.R. Color Control in π—Conjugated Organic Polymers for Use in Electrochromic Devices. *Chem. Rev.* **2010**, *110*, 268–320. [CrossRef]
22. Moliton, A.; Hiorns, R.C. Review of Electronic and Optical Properties of Semiconducting π-Conjugated Polymers: Applications in Optoelectronics. *Polym. Int.* **2004**, *53*, 1397–1412. [CrossRef]
23. Bharti, M.; Singh, A.; Samanta, S.; Aswal, D.K. Conductive Polymers for Thermoelectric Power Generation. *Prog. Mater. Sci.* **2018**, *93*, 270–310. [CrossRef]
24. Fan, Z.; Ouyang, J. Thermoelectric Properties of PEDOT: PSS. *Adv. Electron. Mater.* **2019**, *5*, 1800769. [CrossRef]
25. Guerfi, A.; Trottier, J.; Boyano, I.; De Meatza, I.; Blazquez, J.; Brewer, S.; Ryder, K.; Vijh, A.; Zaghib, K. High cycling stability of zinc-anode/conducting polymer rechargeable battery with non-aqueous electrolyte. *J. Power Sources* **2014**, *248*, 1099–1104. [CrossRef]
26. Katz, H.E.; Searson, P.C.; Poehler, T.O. Batteries and Charge Storage Devices Based on Electronically Conducting Polymers. *J. Mater. Res.* **2010**, *25*, 1561–1574. [CrossRef]
27. Ehsani, A.; Shiri, H.M.; Kowsari, E.; Safari, R.; Torabian, J.; Hajghani, S. High performance electrochemical pseudocapacitors from ionic liquid assisted electrochemically synthesized p-type conductive polymer. *J. Colloid Interface Sci.* **2017**, *490*, 91–96. [CrossRef]
28. Kao, P.; Best, A.S. Conducting-polymer-based supercapacitor devices and electrodes. *J. Power Sources* **2011**, *196*, 1–12.
29. Lee, J.; Kang, H.; Kee, S.; Lee, S.H.; Jeong, S.Y.; Kim, G.; Kim, J.; Hong, S.; Back, H.; Lee, K. Long-Term Stable Recombination Layer for Tandem Polymer Solar Cells Using Self-Doped Conducting Polymers. *ACS Appl. Mater. Interfaces* **2016**, *8*, 6144–6151. [CrossRef] [PubMed]
30. Mengistie, D.A.; Ibraheem, M.A.; Wang, P.C.; Chu, C.W. Highly conductive PEDOT: PSS treated with formic acid for ITO-free polymer solar cells. *ACS Appl. Mater. Interfaces* **2014**, *6*, 2292–2299. [CrossRef]
31. Zhang, J.; Hao, Y.; Yang, L.; Mohammadi, H.; Vlachopoulos, N.; Sun, L.; Hagfeldt, A.; Sheibani, E. Electrochemically polymerized poly (3, 4-phenylenedioxythiophene) as efficient and transparent counter electrode for dye sensitized solar cells. *Electrochim. Acta* **2019**, *300*, 482–488. [CrossRef]
32. Boudreault, P.L.T.; Najari, A.; Leclerc, M. Processable Low-Bandgap Polymers for Photovoltaic Applications. *Chem. Mater.* **2011**, *23*, 456–469. [CrossRef]
33. Kim, Y.H.; Lee, J.; Hofmann, S.; Gather, M.C.; Müller-Meskamp, L.; Leo, K. Achieving high efficiency and improved stability in ITO free transparent organic light-emitting diodes with conductive polymer electrodes. *Adv. Funct. Mater.* **2013**, *23*, 3763–3769. [CrossRef]
34. Bhuvana, K.P.; Joseph Bensingh, R.; Abdul Kader, M.; Nayak, S.K. Polymer Light Emitting Diodes: Materials, Technology and Device. *Polym. Plast. Technol. Eng.* **2018**, *57*, 1784–1800. [CrossRef]
35. Dutta, K.; Das, S.; Rana, D.; Kundu, P.P. Enhancements of Catalyst Distribution and Functioning Upon Utilization of Conducting Polymers as Supporting Matrices in DMFCs: A Review. *Polym. Rev.* **2015**, *55*, 1–56. [CrossRef]

36. Baldissera, A.F.; Freitas, D.B.; Ferreira, C.A. Electrochemical impedance spectroscopy investigation of chlorinated rubber-based coatings containing polyaniline as anticorrosion agent. *Mater. Corros.* **2010**, *61*, 790–801. [CrossRef]
37. Deshpande, P.P.; Jadhav, N.G.; Gelling, V.J.; Sazou, D. Conducting polymers for corrosion protection: A review. *J. Coat. Techn. Res.* **2014**, *11*, 473–494. [CrossRef]
38. Soganci, T.; Gumusay, O.; Soyleyici, H.C.; Ak, M. Synthesis of highly branched conducting polymer architecture for electrochromic applications. *Polymer* **2018**, *134*, 187–195. [CrossRef]
39. Pagès, H.; Topart, P.; Lemordant, D. Wide band electrochromic displays based on thin conducting polymer films. *Electrochim. Acta* **2001**, *46*, 2137–2143. [CrossRef]
40. Barnes, A.; Despotakis, A.; Wong, T.C.P.; Anderson, A.P.; Chambers, B.; Wright, P.V. Towards a 'smart window' for microwave applications. *Smart Mater. Struct.* **1998**, *7*, 752. [CrossRef]
41. Barus, D.A.; Sebayang, K.; Ginting, J.; Ginting, R.T. Effect of Chemical Treatment on Conducting Polymer for Flexible Smart Window Application. *J. Phys. Conf. Ser.* **2018**, *1116*, 032006. [CrossRef]
42. Wang, M.; Wang, X.; Moni, P.; Liu, A.; Kim, D.H.; Jo, W.J.; Sojoudi, H.; Gleason, K.K. CVD Polymers for Devices and Device Fabrication. *Adv. Mater.* **2017**, *29*, 1604606. [CrossRef]
43. Park, C.S.; Kim, D.H.; Shin, B.J.; Tae, H.S. Synthesis and Characterization of Nanofibrous Polyaniline Thin Film Prepared by Novel Atmospheric Pressure Plasma Polymerization Technique. *Materials* **2016**, *9*, 39. [CrossRef]
44. Liu, C.; Goeckner, M.; Walker, A.V. Plasma polymerization of poly (3,4-ethylenedioxyethene) films: The influence of plasma gas phase chemistry. *J. Vac. Sci. Technol. A* **2017**, *35*, 021302. [CrossRef]
45. Loewe, R.S.; Ewbank, P.C.; Liu, J.; Zhai, L.; Mccullough, R.D. Regioregular, Head-to-Tail Coupled Poly(3-Alkylthiophenes) Made Easy by the GRIM Method: Investigation of the Reaction and the Origin of Regioselectivity. *Macromolecules* **2001**, *34*, 4324–4333. [CrossRef]
46. Tamba, S.; Fuji, K.; Meguro, H.; Okamoto, S.; Tendo, T.; Komobuchi, R.; Sugie, A.; Nishino, T.; Mori, A. Synthesis of High-Molecular-Weight Head-to-Tail-Type Poly(3-Substituted-Thiophene)s by Cross-Coupling Polycondensation with [CpNiCl(NHC)] as a Catalyst. *Chem. Lett.* **2013**, *42*, 281–283. [CrossRef]
47. Malinauskas, A. Chemical deposition of conducting polymers. *Polymer* **2001**, *42*, 3957–3972. [CrossRef]
48. Erdem, E.; Karakisla, M.; Sacak, M. The chemical synthesis of conductive polyaniline doped with dicarboxylic acids. *Eur. Polym. J.* **2004**, *40*, 785–791. [CrossRef]
49. Ramanavicius, A.; Kausaite, A.; Ramanaviciene, A. Polypyrrole-coated glucose oxidase nanoparticles for biosensor design. *Sens. Actuators B* **2005**, *111*, 532–539. [CrossRef]
50. Jha, P.; Koiry, S.P.; Saxena, V.; Veerender, P.; Chauhan, A.K.; Aswal, D.K.; Gupta, S.K. Growth of Free-Standing Polypyrrole Nanosheets at Air/Liquid Interface Using J-Aggregate of Porphyrin Derivative as in-Situ Template. *Macromolecules* **2011**, *44*, 4583–4585. [CrossRef]
51. Heinze, J.; Frontana-Uribe, B.A.; Ludwigs, S. Electrochemistry of Conducting Polymers—Persistent Models and New Concepts. *Chem. Rev.* **2010**, *110*, 4724–4771. [CrossRef]
52. Park, Y.; Jung, J.; Chang, M. Research Progress on Conducting Polymer-Based Biomedical Applications. *Appl. Sci.* **2019**, *9*, 1070. [CrossRef]
53. Nair, S.S.; Mishra, S.K.; Kumar, D. Recent progress in conductive polymeric materials for biomedical applications. *Polym. Adv. Technol.* **2019**, *30*, 2932–2953. [CrossRef]
54. Geetha, S.; Rao, C.R.K.; Vijayan, M.; Trivedi, D.C. Biosensing and drug delivery by polypyrrole. *Anal. Chim. Acta* **2006**, *568*, 119–125. [CrossRef] [PubMed]
55. Boehler, C.; Oberueber, F.; Asplund, M. Tuning drug delivery from conducting polymer films for accurately controlled release of charged molecules. *J. Control. Release* **2019**, *304*, 173–180. [CrossRef]
56. Krukiewicz, K.; Bednarczyk, B.; Turczyn, R.; Zak, J.K. EQCM verification of the concept of drug immobilization and release from conducting polymer matrix. *Electrochim. Acta* **2016**, *212*, 694–700. [CrossRef]
57. Guo, B.; Ma, P.X. Conducting polymers for tissue engineering. *Biomacromolecules* **2018**, *19*, 1764–1782. [CrossRef] [PubMed]
58. Dong, R.; Ma, P.X.; Guo, B. Conductive biomaterials for muscle tissue engineering. *Biomaterials* **2020**, *229*, 119584. [CrossRef]
59. Zarrintaj, P.; Bakhshandeh, B.; Saeb, M.R.; Sefat, F.; Rezaeian, I.; Ganjali, M.R.; Ramakrishna, S.; Mozafari, M. Oligoaniline-based conductive biomaterials for tissue engineering. *Acta Biomater.* **2018**, *72*, 16–34. [CrossRef]

60. Inal, S.; Hama, A.; Ferro, M.; Pitsalidis, C.; Oziat, J.; Iandolo, D.; Pappa, A.M.; Hadida, M.; Huerta, M.; Marchat, D.; et al. Conducting polymer scaffolds for hosting and monitoring 3D cell culture. *Adv. Biosyst.* **2017**, *1*, 1700052. [CrossRef]
61. Lakard, S.; Morrand-Villeneuve, N.; Lesniewska, E.; Lakard, B.; Michel, G.; Herlem, G.; Gharbi, T.; Fahys, B. Synthesis of polymer materials for use as cell culture substrates. *Electrochim. Acta* **2007**, *53*, 1114–1126. [CrossRef]
62. Ateh, D.D.; Navsaria, H.A.; Vadgama, P. Polypyrrole-based conducting polymers and interactions with biological tissues. *J. R. Soc. Interface* **2006**, *3*, 741–752. [CrossRef]
63. He, H.; Zhang, L.; Guan, X.; Cheng, H.; Liu, X.; Yu, S.; Wei, J.; Ouyang, J. Biocompatible conductive polymers with high conductivity and high stretchability. *ACS Appl. Mater. Interfaces* **2019**, *11*, 26185–26193. [CrossRef]
64. Humpolicek, P.; Kasparkova, V.; Pachernik, J.; Stejskal, J.; Bober, P.; Capakova, Z.; Radaszkiewicz, K.A.; Junkar, I.; Lehocky, M. The biocompatibility of polyaniline and polypyrrole: A comparative study of their cytotoxicity, embryotoxicity and impurity profile. *Mat. Sci. Eng. C* **2018**, *91*, 303–310. [CrossRef]
65. Humpolicek, P.; Kasparkova, V.; Saha, P.; Stejskal, J. Biocompatibility of polyaniline. *Synth. Met.* **2012**, *162*, 722–727. [CrossRef]
66. George, P.M.; Lyckman, A.W.; LaVan, D.A.; Hegde, A.; Leung, Y.; Avasare, R.; Testa, C.; Alexander, P.M.; Langer, R.; Sur, M. Fabrication and biocompatibility of polypyrrole implants suitable for neural prosthetics. *Biomaterials* **2005**, *26*, 3511–3519. [CrossRef] [PubMed]
67. Kim, D.M.; Cho, S.J.; Cho, C.H.; Kim, K.B.; Kim, M.Y.; Shim, Y.B. Disposable all-solid-state pH and glucose sensors based on conductivepolymer covered hierarchical AuZn oxide. *Biosens. Bioelectron.* **2016**, *79*, 165–172. [CrossRef] [PubMed]
68. Tuncagil, S.; Ozdemir, C.; Demirkol, D.O.; Timur, S.; Toppare, L. Gold nanoparticle modified conducting polymer of 4-(2,5-di(thiophen-2-yl)-1H-pyrrole-1-l) benzenamine for potential use as a biosensing material. *Food Chem.* **2011**, *127*, 1317–1322. [CrossRef]
69. Kausaite-Minkstimiene, A.; Glumbokaite, L.; Ramanaviciene, A.; Dauskaite, E.; Ramanavicius, A. An Amperometric Glucose Biosensor Based on Poly (Pyrrole-2-Carboxylic Acid)/Glucose Oxidase Biocomposite. *Electroanalysis* **2018**, *30*, 1642–1652. [CrossRef]
70. Gaikwad, P.; Shirale, D.; Gade, V.; Savale, P.; Kharat, H.; Kakde, K.; Shirsat, M. Immobilization of GOD on electrochemically synthesized PANI film by cross-linking via glutaraldehyde for determination of glucose. *Int. J. Electrochem. Sci.* **2006**, *1*, 425–434.
71. Lakard, B.; Herlem, G.; Lakard, S.; Antoniou, A.; Fahys, B. Urea potentiometric biosensor based on modified electrodes with urease immobilized on polyethylenimine films. *Biosens. Bioelectron.* **2004**, *19*, 1641–1647. [CrossRef]
72. Magnin, D.; Callegari, V.; Matefi-Tempfli, S.; Matefi-Tempfli, M.; Glinel, K.; Jonas, A.M.; Demoustier-Champagne, S. Functionalization of Magnetic Nanowires by Charged Biopolymers. *Biomacromolecules* **2008**, *9*, 2517–2522. [CrossRef]
73. Xue, H.G.; Shen, Z.Q.; Li, Y.F. Polyaniline-polyisoprene composite film based glucose biosensor with high permselectivity. *Synth. Met.* **2001**, *124*, 345–349. [CrossRef]
74. Molino, P.J.; Higgins, M.J.; Innis, P.C.; Kapsa, R.M.I.; Wallace, G.G. Fibronectin and bovine serum albumin adsorption and conformational dynamics on inherently conducting polymers: A QCM-D study. *Langmuir* **2012**, *28*, 8433–8445. [CrossRef] [PubMed]
75. Lakard, B.; Magnin, D.; Deschaume, O.; Vanlancker, G.; Glinel, K.; Demoustier-Champagne, S.; Nysten, B.; Jonas, A.M.; Bertrand, P.; Yunus, S. Urea potentiometric enzymatic biosensor based on charged biopolymers and electrodeposited polyaniline. *Biosens. Bioelectron.* **2011**, *26*, 4139–4145. [CrossRef] [PubMed]
76. Yang, G.; Kampstra, K.L.; Abidian, M.R. High performance conducting polymer nanofiber biosensors for detection of biomolecules. *Adv. Mater.* **2014**, *26*, 4954–4960. [CrossRef] [PubMed]
77. Soares, J.C.; Brisolari, A.; da Cruz Rodrigues, V.; Sanches, E.A.; Gonçalves, D. Amperometric urea biosensors based on the entrapment of urease in polypyrrole films. *React. Funct. Polym.* **2012**, *72*, 148–152. [CrossRef]
78. Minett, A.I.; Barisci, J.N.; Wallace, G.G. Immobilisation of anti-Listeria in a polypyrrole film. *React. Funct. Polym.* **2002**, *53*, 217–227. [CrossRef]
79. Mandli, J.; Amine, A. Impedimetric genosensor for miRNA-34a detection in cell lysates using polypyrrole. *J. Solid State Electrochem.* **2018**, *22*, 1007–1014. [CrossRef]
80. Ramanavicius, A.; Ramanaviciene, A.; Malinauskas, A. Electrochemical sensors based on conducting polymer-polypyrrole. *Electrochim. Acta* **2006**, *51*, 6025–6037. [CrossRef]

81. Adeloju, S.B.; Moline, A.N. Fabrication of ultra-thin polypyrrole–glucose oxidase film from supporting electrolyte-free monomer solution for potentiometric biosensing of glucose. *Biosens. Bioelectron.* **2001**, *16*, 133–139. [CrossRef]
82. Leite, C.; Lakard, B.; Hihn, J.Y.; del Campo, F.J.; Lupu, S. Use of sinusoidal voltages with fixed frequency in the preparation of tyrosinase based electrochemical biosensors for dopamineelectroanalysis. *Sens. Actuators B* **2017**, *240*, 801–809. [CrossRef]
83. Clark, L.C. Monitor and control of blood and tissue oxygen tensions. *Trans. Am. Soc. Artif. Intern. Organs* **1956**, *2*, 41–48.
84. Forzani, E.S.; Zhang, H.; Nagahara, L.A.; Amlani, I.; Tsui, R.; Tao, N. A Conducting Polymer Nanojunction Sensor for Glucose Detection. *Nano Lett.* **2004**, *4*, 1785–1788. [CrossRef]
85. Chauhan, N.; Chawla, S.; Pundir, C.S.; Jain, U. An electrochemical sensor for detection of neurotransmitter-acetylcholine using metal nanoparticles, 2D material and conducting polymer modified electrode. *Biosens. Bioelectron.* **2017**, *89*, 377–383. [CrossRef]
86. Uwaya, G.E.; Fayemi, O.E. Electrochemical detection of serotonin in banana at green mediated PPy/Fe_3O_4 NPs nanocomposites modified electrodes. *Sens. Bio Sens. Res.* **2020**, *28*, 100338. [CrossRef]
87. Hamid, H.H.; Harb, M.E.; Elshaer, A.M.; Erahim, S.; Soliman, M.M. Electrochemical Preparation and Electrical Characterization of Polyaniline as a Sensitive Biosensor. *Microsyst. Technol.* **2018**, *24*, 1775–1781. [CrossRef]
88. Bahadir, E.B.; Sezgintürk, M.K. A review on impedimetric biosensors. *Artif. Cells Nanomed Biotechn.* **2016**, *44*, 248–262. [CrossRef]
89. He, S.; Yuan, Y.; Nag, A.; Feng, S.; Afsarimanesh, N.; Han, T.; Mukhopadhyay, S.C.; Organ, D.R. A Review on the Use of Impedimetric Sensors for the Inspection of Food Quality. *Int. J. Environ. Res. Public Health* **2020**, *17*, 5220. [CrossRef] [PubMed]
90. Leva-Bueno, J.; Peyman, S.A.; Millner, P.A. A review on impedimetric immunosensors for pathogen and biomarker detection. *Med. Microbiol. Immunol.* **2020**, *209*, 343–362. [CrossRef]
91. Grant, S.; Davis, F.; Law, K.A.; Barton, A.C.; Collyer, S.D.; Higson, S.P.J.; Gibson, T.D. Label-free and reversible immunosensor based upon an ac impedance interrogation protocol. *Anal. Chim. Acta* **2005**, *537*, 163–168. [CrossRef]
92. Aydin, E.B.; Aydin, M.; Sezgintürk, M.K. Highly sensitive electrochemical immunosensor based on polythiophene polymer with densely populated carboxyl groups as immobilization matrix for detection of interleukin 1β in human serum and saliva. *Sens. Actuators B* **2018**, *270*, 18–27. [CrossRef]
93. Taleat, Z.; Ravalli, A.; Mazloum-Ardakami, M.; Marrazza, G. CA 125 Immunosensor Based on Poly-Anthranilic Acid Modified Screen-Printed Electrodes. *Electroanalysis* **2013**, *25*, 269–277. [CrossRef]
94. Jugovic, B.; Grgur, B.; Antov, M.; Knezevic-Jugovic, Z.; Stevanovic, J.; Gvozdenovic, M. Polypyrrole-based Enzyme Electrode with Immobilized Glucose Oxidase for Electrochemical Determination of Glucose. *Int. J. Electrochem. Sci.* **2016**, *11*, 1152–1161.
95. Gvozdenovic, M.M.; Jugovic, B.Z.; Bezbradica, D.I.; Antov, M.G.; Knezevic-Jugovic, Z.D.; Grgur, B.N. Electrochemical determination of glucose using polyaniline electrode modified by glucose oxidase. *Food Chem.* **2011**, *124*, 396–400. [CrossRef]
96. Lai, J.; Yi, Y.; Zhu, P.; Shen, J.; Wu, K.; Zhang, L.; Liu, J. Polyaniline-based glucose biosensor: A review. *J. Electroanal. Chem.* **2016**, *782*, 138–153. [CrossRef]
97. Singh, M.; Kathuroju, P.K.; Jampana, N. Polypyrrole based amperometric glucose biosensors. *Sens. Actuators B* **2009**, *143*, 430–443. [CrossRef]
98. Lau, K.T.; de Fortescu, S.A.L.; Murphy, L.J.; Slater, J.M. Disposable glucose sensors for flow injection analysis using substituted 1,4-benzoquinonemediators. *Electroanalysis* **2003**, *15*, 975–981. [CrossRef]
99. Qiu, J.D.; Zhou, W.M.; Guo, J.; Wang, R.; Liang, R.P. Amperometric sensor based on ferrocene-modified multiwalled carbon nanotube nanocomposites as electron mediator for the determination of glucose. *Anal. Biochem.* **2009**, *385*, 264–269. [CrossRef]
100. Jian, L.; Shanhui, S.; Changchun, L.; Shoushui, W. An amperometric glucose biosensor based on a screen-printed electrode and Os-complex mediator for flow injection analysis. *Measurement* **2011**, *44*, 1878–1883.

101. Shrestha, B.K.; Ahmad, R.; Mousa, H.M.; Kim, I.G.; Kim, J.I.; Neupane, M.P.; Park, C.H.; Kim, C.S. High-performance glucose biosensor based on chitosane-glucose oxidase immobilized polypyrrole/Nafion/functionalized multi-walled carbon nanotubes bio-nanohybrid film. *J. Colloid Interf. Sci.* **2016**, *482*, 39–47. [CrossRef]
102. Shrestha, B.K.; Ahmad, R.; Shrestha, S.; Park, C.H.; Kim, C.S. Globular Shaped Polypyrrole Doped Well-Dispersed Functionalized Multiwall Carbon Nanotubes/Nafion Composite for Enzymatic Glucose Biosensor Application. *Sci. Rep.* **2017**, *7*, 16191. [CrossRef]
103. Chen, X.; Chen, Z.; Tian, R.; Yan, W.; Yao, C. Glucose biosensor based on three dimensional ordered macroporous self-doped polyaniline/Prussian blue bicomponent film. *Anal. Chim. Acta* **2012**, *723*, 94–100. [CrossRef]
104. Chowdhury, A.D.; Gangopadhyay, R.; De, A. Highly sensitive electrochemical biosensor for glucose, DNA and protein using gold-polyaniline nanocomposites as a common matrix. *Sens. Actuators B* **2014**, *190*, 348–356. [CrossRef]
105. Mazeiko, V.; Kausaite-Minkstimiene, A.; Ramanaviciene, A.; Balevicius, Z.; Ramanavicius, A. Gold Nanoparticle and Conducting Polymer—Polyaniline—Based Nanocomposites for Glucose Biosensor Design. *Sens. Actuators B* **2013**, *189*, 187–193. [CrossRef]
106. Zhai, D.; Liu, B.; Shi, Y.; Pan, L.; Wang, Y.; Li, Y.; Zhang, R.; Yu, G. Highly sensitive glucose sensor based on Pt nanoparticle/polyaniline hydrogel heterostructures. *ACS Nano* **2013**, *7*, 3540–3546. [CrossRef] [PubMed]
107. Zhong, H.; Yuan, R.; Chai, Y.; Li, W.; Zhong, X.; Zhang, Y. In situ chemo-synthesized multi-wall carbon nanotube-conductive polyaniline nanocomposites: Characterization and application for a glucose amperometric biosensor. *Talanta* **2011**, *85*, 104–111. [CrossRef]
108. Jimenez-Fierrez, F.; Gonzalez-Sanchez, M.I.; Jimenez-Perez, R.; Iniesta, J.; Valero, E. Glucose Biosensor Based on Disposable Activated Carbon Electrodes Modified with Platinum Nanoparticles Electrodeposited on Poly(Azure A). *Sensors* **2020**, *20*, 4489. [CrossRef]
109. Djaalab, E.; El Hadi Samar, M.; Zougar, S.; Kherrat, R. Electrochemical Biosensor for the Determination of Amlodipine Besylate Based on Gelatin-Polyaniline Iron Oxide Biocomposite Film. *Catalysts* **2018**, *8*, 233. [CrossRef]
110. Zhuang, X.; Tian, C.; Luan, F.; Wu, X.; Chen, L. One-step electrochemical fabrication of a nickel oxide nanoparticle/polyaniline nanowire/graphene oxide hybrid on a glassy carbon electrode for use as a non-enzymatic glucose biosensor. *RSC Adv.* **2016**, *6*, 92541–92546. [CrossRef]
111. Wang, Z.; Liu, S.; Wu, P.; Cai, C. Detection of Glucose Based on Direct Electron Transfer Reaction of Glucose Oxidase Immobilized on Highly Ordered Polyaniline Nanotubes. *Anal. Chem.* **2009**, *81*, 1638–1645. [CrossRef]
112. Xu, G.; Adeloju, S.B.; Wu, Y.; Zhang, X. Modification of polypyrrole nanowires array with platinum nanoparticles and glucose oxidase for fabrication of a novel glucose biosensor. *Anal. Chim. Acta* **2012**, *755*, 100–107. [CrossRef]
113. Komathi, S.; Gopalan, A.I.; Muthuchamy, N.; Lee, K.P. Polyaniline nanoflowers grafted onto nanodiamonds via a soft template-guided secondary nucleation process for high-performance glucose sensing. *RSC Adv.* **2017**, *7*, 15342–15351. [CrossRef]
114. Ramanavicius, A.; Habermüller, K.; Csöregi, E.; Laurinavicius, V.; Schuhmann, W. Polypyrrole-Entrapped Quinohemoprotein Alcohol Dehydrogenase. Evidence for Direct Electron Transfer via Conducting-Polymer Chains. *Anal. Chem.* **1999**, *71*, 3581–3586. [CrossRef]
115. Bollela, P.; Gorton, L.; Antiochia, R. Direct Electron Transfer of Dehydrogenases for Development of 3rd Generation Biosensors and Enzymatic Fuel Cells. *Sensors* **2018**, *18*, 1319. [CrossRef] [PubMed]
116. Oztekin, Y.; Ramanaviciene, A.; Yazicigil, Z.; Solak, A.O.; Ramanavicius, A. Direct electron transfer from glucose oxidase immobilized on polyphenanthroline-modified glassy carbon electrode. *Biosens. Bioelectron.* **2011**, *26*, 2541–2546. [CrossRef] [PubMed]
117. Bagdziunas, G.; Palinauskas, D. Poly (9H-carbazole) as a Organic Semiconductor for Enzymatic and Non-Enzymatic Glucose Sensors. *Biosensors* **2020**, *10*, 104. [CrossRef] [PubMed]
118. Grennan, K.; Strachan, G.; Porter, A.J.; Killard, A.J.; Smyth, M.R. Atrazine analysis using an amperometric immunosensor based on single-chain antibody fragments and regeneration-free multi-calibrant measurement. *Anal. Chim. Acta* **2003**, *500*, 287–298. [CrossRef]

119. Darain, F.; Park, D.S.; Park, J.S.; Shim, Y.B. Development of an immunosensor for the detection of vitellogenin using impedance spectroscopy. *Biosens. Bioelectron.* **2004**, *19*, 1245–1252. [CrossRef]
120. Wang, H.; Ma, F. A cascade reaction signal-amplified amperometric immunosensor platform for ultrasensitive detection of tumour marker. *Sens. Actuators B* **2018**, *254*, 642–647. [CrossRef]
121. Zhao, L.; Ma, Z. Facile Synthesis of Polyaniline-Polythionine Redox Hydrogel: Conductive, Antifouling and Enzyme-Linked Material for Ultrasensitive Label-Free Amperometric Immunosensor toward Carcinoma Antigen-125. *Anal. Chim. Acta* **2018**, *997*, 60–66. [CrossRef]
122. Shaikh, M.O.; Srikanth, B.; Zhu, P.Y.; Chuang, C.H. Impedimetric Immunosensor Utilizing Polyaniline/Gold Nanocomposite-Modified Screen-Printed Electrodes for Early Detection of Chronic Kidney Disease. *Sensors* **2019**, *19*, 3990. [CrossRef]
123. Liu, S.; Ma, Y.; Cui, M.; Luo, X. Enhanced Electrochemical Biosensing of Alpha-Fetoprotein Based on Three-Dimensional Macroporous Conducting Polymer Polyaniline. *Sens. Actuators B* **2018**, *255*, 2568–2574. [CrossRef]
124. Dutta, S.; Chowdhury, A.D.; Biswas, S.; Park, E.Y.; Agnihotri, N.; De, A.; De, S. Development of an effective electrochemical platform for highly sensitive DNA detection using MoS_2—polyaniline nanocomposites. *Biochem. Eng. J.* **2018**, *140*, 130–139. [CrossRef]
125. Eguiluz, K.I.V.; Salazar-Banda, G.R.; Elizabeth, M.; Huacca, F.; Alberice, J.V.; Carrilho, E.; Machado, S.A.S.; Avaca, L.A. Sequence-specific electrochemical detection of Alicyclobacillus acidoterrestris DNA using electroconductive polymer-modified fluorine tin oxide electrodes. *Analyst* **2009**, *134*, 314–319. [CrossRef] [PubMed]
126. Tlili, C.; Jaffrezic-Renault, N.J.; Martelet, C.; Korri-Youssoufi, H. Direct electrochemical probing of DNA hybridization on oligonucleotide-functionalized polypyrrole. *Mater. Sci. Eng. C* **2008**, *28*, 848–854. [CrossRef]
127. Wilson, J.; Radhakrishnan, S.; Sumathi, C.; Dharuman, V. Polypyrrole-polyaniline-Au (PPy-PANi-Au) nano composite films for label-free electrochemical DNA sensing. *Sens. Actuators B* **2012**, *171*, 216–222. [CrossRef]
128. Radhakrishnan, S.; Sumathi, C.; Umar, A.; Kim, S.J.; Wilson, J.; Dharuman, V. Polypyrrole-poly (3,4-ethylenedioxythiophene)-Ag (PPy-PEDOT-Ag) nanocomposite films for label-free electrochemical DNA sensing. *Biosens. Bioelectron.* **2013**, *47*, 133–140. [CrossRef] [PubMed]
129. Xu, Y.; Ye, X.; Yang, L.; He, P.; Fang, Y. Impedance DNA biosensor using electropolymerized polypyrrole/multiwalled carbon nanotubes modified electrode. *Electroanalysis* **2006**, *18*, 1471–1478. [CrossRef]
130. Chang, H.; Yuan, Y.; Shi, N.; Guan, Y. Electrochemical DNA biosensor based on conducting polyaniline nanotube Array. *Anal. Chem.* **2007**, *79*, 5111–5115. [CrossRef]
131. Zhu, N.; Chang, Z.; He, P.; Fang, Y. Electrochemically fabricated polyaniline nanowire-modified electrode for voltammetric detection of DNA hybridization. *Electrochim. Acta* **2006**, *51*, 3758–3762. [CrossRef]
132. Du, M.; Yang, T.; Li, X.; Jiao, K. Fabrication of DNA/graphene/polyaniline nanocomplex for label-free voltammetric detection of DNA hybridization. *Talanta* **2012**, *88*, 439–444. [CrossRef]
133. Mohamad, F.S.; Zaid, M.H.M.; Abdullah, J.; Zawawi, R.M.; Lim, H.N.; Sulaiman, Y.; Rahman, N.A. Synthesis and Characterization of Polyaniline/Graphene Composite Nanofiber and Its Application as an Electrochemical DNA Biosensor for the Detection of Mycobacterium tuberculosis. *Sensors* **2017**, *17*, 2789. [CrossRef]
134. Ding, L.; Du, D.; Zhang, X.; Ju, H. Trends in Cell-Based Electrochemical Biosensors. *Curr. Med. Chem.* **2008**, *15*, 3160–3170. [CrossRef] [PubMed]
135. Wei, Y.; Mo, X.; Zhang, P.; Li, Y.; Liao, J.; Li, Y.; Zhang, J.; Ning, C.; Wang, S.; Deng, X.; et al. Directing Stem Cell Differentiation via Electrochemical Reversible Switching between Nanotubes and Nanotips of Polypyrrole Array. *ACS Nano* **2017**, *11*, 5915–5924. [CrossRef]
136. Lakard, B.; Ploux, L.; Anselme, K.; Lallemand, F.; Lakard, S.; Nardin, M.; Hihn, J.Y. Effect of ultrasounds on the electrochemical synthesis of polypyrrole. Application to the adhesion and growth of biological cells. *Bioelectrochemistry* **2009**, *75*, 148–157. [CrossRef] [PubMed]
137. El-Said, W.A.; Yea, C.H.; Choi, J.W.; Kwon, I.K. Ultrathin polyaniline film coated on an indium-tin oxide cell-based chip for study of anticancer effect. *Thin Solid Film* **2009**, *518*, 661–667. [CrossRef]
138. Lee, J.Y.; Lee, J.W.; Schmidt, C.E. Neuroactive conducting scaffolds: Nerve growth factor conjugation on active ester-functionalized polypyrrole. *J. R. Soc. Interface* **2009**, *6*, 801–810. [CrossRef]
139. Nguyen-Vu, T.D.B.; Chen, H.; Cassell, A.M.; Andrews, R.; Meyyappan, M.; Li, J. Vertically Aligned Carbon Nanofiber Arrays: An Advance toward Electrical-Neural Interfaces. *Small* **2006**, *2*, 89–94. [CrossRef]

140. El-Said, W.A.; Yea, C.H.; Jung, M.; Kim, H.C.; Choi, J.W. Analysis of effect of nanoporous alumina substrate coated with polypyrrole nanowire on cell morphology based on AFM topography. *Ultramicroscopy* **2010**, *110*, 676–681. [CrossRef]
141. Apetrei, R.M.; Carac, G.; Bahrim, G.; Camurlu, P. Glucose biosensor based on whole cells of Aspergillus niger MIUG 34 coated with polypyrrole. *J. Biotechnol.* **2017**, *256*, S55–S56. [CrossRef]
142. Cevik, E.; Cerit, A.; Tombuloglu, H.; Sabit, H.; Yildiz, H.B. Electrochemical Glucose Biosensors: Whole Cell Microbial and Enzymatic Determination based on 10-(4H-dithiyeno [3,2-b:2′,3′-d]pyroll-4-il) decan-1-amine Interfaces Glassy Carbon Electrodes. *Anal. Lett.* **2019**, *52*, 1138–1152. [CrossRef]
143. Baskurt, E.; Ekiz, F.; Demirkol, D.O.; Timur, S.; Toppare, L. A conducting polymer with benzothiadiazole unit: Cell based biosensing applications and adhesion properties. *Colloids Surf. B* **2012**, *97*, 13–18. [CrossRef]
144. Guler, E.; Soylevici, H.C.; Demirkol, D.O.; Ak, M.; Timur, S. A novel functional conducting polymer as an immobilization platform. *Mat. Sci. Eng. C* **2014**, *40*, 148–156. [CrossRef] [PubMed]
145. Tuncagil, S.; Odaci, D.; Yildiz, E.; Timur, S.; Toppare, L. Design of a microbial sensor using conducting polymer of 4-(2,5-di(thiophen-2-yl)-1H-pyrrole-1-l) benzenamine. *Sens. Actuators B* **2009**, *137*, 42–47. [CrossRef]
146. Jha, S.K.; Kanungo, M.; Nath, A.; D'Souza, S.F.D. Entrapment of live microbial cells in electropolymerized polyaniline and their use as urea biosensor. *Biosens. Bioelectron.* **2009**, *24*, 2637–2642. [CrossRef] [PubMed]
147. Ratautaite, V.; Plausinaitis, D.; Baleviciute, I.; Mikoliunaite, L.; Ramanaviciene, A.; Ramanavicius, A. Characterization of caffeine-imprinted polypyrrole by a quartz crystalmicrobalance and electrochemical impedance spectroscopy. *Sens. Actuators B* **2015**, *212*, 63–71. [CrossRef]
148. Xing, X.; Liu, S.; Yu, J.; Lian, W.; Huan, J. Electrochemical sensor based on molecularly imprinted film at polypyrrole-sulfonated graphene/hyaluronic acid-multiwalled carbon nanotubes modified electrode for determination of tryptamine. *Biosens. Bioelectron.* **2012**, *31*, 277–283. [CrossRef]
149. Zhou, H.; Xu, G.; Zhu, A.; Zhao, Z.; Ren, C.; Nie, L.; Kan, X. A multiporous electrochemical sensor for epinephrine recognition and detection based on molecularly imprinted polypyrrole. *RSC Advances* **2012**, *2*, 7803–7808. [CrossRef]
150. Radi, A.E.; El-Naggar, A.E.; Nassef, H.M. Determination of coccidiostat clopidol on an electropolymerized-molecularly imprinted polypyrrole polymer modified screen printed carbon electrode. *Anal. Methods* **2014**, *6*, 7967–7972. [CrossRef]
151. Lai, Y.X.; Zhang, C.X.; Deng, Y.; Yang, G.J.; Li, S.; Tang, C.L.; He, N.Y. A novel alpha-fetoprotein-MIP immunosensor based on AuNPs/PTh modified glass carbon electrode. *Chin. Chem. Lett.* **2019**, *30*, 160–162. [CrossRef]
152. Karami, P.; Bagheri, H.; Johari-Ahar, M.; Khoshsafar, H.; Arduini, F.; Afkhami, A. Dual-modality impedimetric immunosensor for early detection of prostate-specific antigen and myoglobin markers based on antibody-molecularly imprinted polymer. *Talanta* **2019**, *202*, 111–122. [CrossRef]

© 2020 by the author. Licensee MDPI, Basel, Switzerland. This article is an open access article distributed under the terms and conditions of the Creative Commons Attribution (CC BY) license (http://creativecommons.org/licenses/by/4.0/).

Article

Self-Assembled MoS₂/ssDNA Nanostructures for the Capacitive Aptasensing of Acetamiprid Insecticide

Maroua Hamami [1,2], Noureddine Raouafi [2,*] and Hafsa Korri-Youssoufi [2,*]

1. Université Paris-Saclay, CNRS, Institut de Chimie Moléculaire et des Matériaux d'Orsay (ICMMO), ECBB, Bât 420, 2 Rue du Doyen Georges Poitou, 91400 Orsay, France; maroua.hamami@universite-paris-saclay.fr
2. Faculté des Sciences de Tunis El Manar, Laboratoire de Chimie Analytique et Electrochimie (LR99ES15), Campus Universitaire de Tunis El Manar, Université de Tunis El Manar, Tunis El–Manar 2092, Tunisia
* Correspondence: hafsa.korri-youssoufi@universite-paris-saclay.fr (N.R.); noureddine.raouafi@fst.utm.tn (H.K.-Y.)

Featured Application: Authors are encouraged to provide a concise description of the specific application or a potential application of the work. This section is not mandatory.

Abstract: The aim of this work is to detect acetamiprid using electrochemical capacitance spectroscopy, which is widely used as a pesticide in agriculture and is harmful to humans. We have designed aptasensing platform based on the adsorption of a DNA aptamer on lipoic acid-modified MoS₂ nano-sheets. The biosensor takes advantage of the high affinity of single-stranded DNA sequences to MoS₂ nano-sheets. The stability of DNA on MoS₂ nano-sheets is assured by covalent attachment to lipoic acid that forms self-assembled layer on MoS₂ surface. The biosensor exhibits excellent capacitance performances owing to its large effective surface area making it interesting material for capacitive transduction system. The impedance-derived capacitance varies with the increasing concentrations of acetamiprid that can be attributed to the aptamer desorption from the MoS₂ nanosheets facilitating ion diffusion into MoS₂ interlayers. The developed device showed high analytical performances for acetamiprid detection on electrochemical impedance spectroscopy EIS-derived capacitance variation and high selectivity toward the target in presence of other pesticides. Real sample analysis of food stuff such as tomatoes is demonstrated which open the way to their use for monitoring of food contaminants by tailoring the aptamer.

Keywords: aptasensor; MoS₂; pesticide; neonicotinoid; capacitance

Citation: Hamami, M.; Raouafi, N.; Korri-Youssoufi, H. Self-Assembled MoS₂/ssDNA Nanostructures for the Capacitive Aptasensing of Acetamiprid Insecticide. *Appl. Sci.* 2021, 11, 1382. https://doi.org/10.3390/app11041382

Academic Editor: Tae Hyun Kim
Received: 17 December 2020
Accepted: 26 January 2021
Published: 3 February 2021

Publisher's Note: MDPI stays neutral with regard to jurisdictional claims in published maps and institutional affiliations.

Copyright: © 2021 by the authors. Licensee MDPI, Basel, Switzerland. This article is an open access article distributed under the terms and conditions of the Creative Commons Attribution (CC BY) license (https://creativecommons.org/licenses/by/4.0/).

1. Introduction

Neonicotinoid acetamiprid is considered as one of the most efficient neuro-active insecticides, so large amounts of acetamiprid are routinely used in agriculture for treatment of numerous pests. Due to its abuse of use, the threat of pesticide residues to human health and environmental pollution are a real public concern [1]. For instance, it has been reported that acetamiprid could affect human peripheral blood lymphocytes and cause DNA damages [2] and its accumulation in agricultural products is a serious threat for human beings [3]. Thus, the detection and monitoring of acetamiprid levels in foodstuffs are of great interest for public health safety. The United States Environmental Protection Agency (EPA) and the European Food Safety Authority have set the maximum residue level (MRL) of acetamiprid from 0.01 to 3 ppm depending on the nature of the vegetables and the legislation is regularly revised regarding this MRL [4]. However, detecting pesticides at these levels is still challenging. Therefore, the development of reliable, sensitive, direct, and fast analytical methods for the acetamiprid analysis in fresh products is of paramount importance. For this purpose, various analytical methods are available such as enzyme-linked immunosorbent assays [5], gas chromatography [6], high-performance liquid chromatography [7], and gas chromatography–mass spectrometry [8]. Nevertheless, these techniques have some limitations like high equipment costs, suitable only for laboratory analysis, time-consuming due to sample preparation and require trained technicians to operate them. Hence, the development of simple, cost-effective, sensitive, and portable alternatives is necessary for the fast detection of acetamiprid in environment and agricultural field to reduce the risk of public health.

Over the last decade, electrochemical biosensors have gained increasing interest in the field of food monitoring due to their excellent properties and remarkable analytical performances [9]. Especially, oligonucleotide-based electrochemical biosensors are increasingly applied for sensitive detection of pesticide residues [10]. Aptamers are single-stranded oligonucleotides considered as excellent molecular probes, which are endowed with high affinity for various target substances, including small molecules [11], viruses [12] proteins [13], and cells [14]. In an attempt to develop electrochemical aptasensors, several acetamiprid sensing approaches based on the aptamer-target affinity have been developed. For instance, Shi et al. [15] recently reported a dual signal amplification strategy for the aptasensing of acetamiprid using reduced graphene oxide and silver nanoparticles. The electrical signal recorded by cyclic voltammetry was significantly improved after the immobilization of Prussian blue-gold nanoparticles as a catalyst for the redox reaction. In another competition-based strategy, silica nanoparticles modified dsDNA formed by the perfect match of the aptamer and a complementary sequence intercalated with methylene blue (MB) as a redox probe are used. In the presence of acetamiprid, the high affinity of aptamer toward the target induced the release of MB that is detected electrochemically using an unmodified gold electrode [16]. In the other work, Fei et al. [17] did not use a redox probe and proposed a label free impedimetric aptasensor based on complex composites of gold nanoparticles (AuNPs) decorated MWCNTs and reduced graphene oxide nanoribbons to detect femtomolar levels of the target. The aforementioned studies showed the use of conventional electrochemical techniques such as cyclic voltammetry and electrochemical impedance spectroscopy. Recent efforts focused to optimize and to enhance the signal sensitivity of these techniques and particularly of electrochemical impedance spectroscopy that needs theoretical modeling by an equivalent circuit. Accordingly, Santos et al. [18] have been working intensively on the electrochemical capacitance spectroscopy (ECS) or impedance-derived capacitance approach as a better alternative. They showed that when the system is non-faradaic or faradaic regime with an external redox probe (in solution), the electrochemical capacitance can be represented by the double layer capacitance (C_{dl}). In the case of a redox marker attached to the electrode surface, a pseudo-capacitance called redox capacitance (C_r) that depends on the density-of-states of the confined redox marker was considered instead the C_{dl} [19]. This can be explained by the fact that measuring electrochemical impedance spectroscopy (EIS) at the half-wave potential of the confined redox system, the contribution of C_r is enhanced and the C_{dl} remains almost constant and therefore its contribution can be neglected [20]. The C_r element corresponds to the diameter of the semicircle in Nyquist capacitive plots and its value is obtained by converting the Nyquist EIS plots. This technique has been successfully adapted for various sensing systems [21]. According to Fernandes et al. [22], the redox capacitance of a faradaic probe confined within a biological film on the surface, is sensitive to the whole system thus can be correlated to the analyte concentration when the biological film is capable of recognizing the target with high specificity. This implies that EIS-derived capacitance signal is based on the charging signal that is generated from the activity of electroactive tethered groups, which is related to its electrostatic environment.

The aim of this work was to report a new strategy based on the use of redox-active nanomaterials that will not only replace the confined redox probe but also will enhance the biosensor performances by using their capacitance properties. Particularly, two-dimensional (2D) nanomaterials have shown interesting properties that helped to improve the sensitivity and analytical performances of the developed biosensors [23]. In fact, MoS_2 nanosheets attracted increasing interests for its excellent capacitive properties [24]. Their sheet-like morphology provides large surface area for charge storage that can potentially occur via faradaic charge transfer process on the Mo(+IV) metal cation. The Mo center presenting a range of oxidation states from +2 to +6 can exhibit pseudo-capacitance [25]. Hence, 2D MoS$_2$ is an excellent choice for electrochemical capacitance spectroscopy application. Furthermore, the MoS$_2$ possesses an ultrathin plane structure of atomic thickness making it sensitive to the surrounding environment [26]; thus, an interaction with a target biomolecule can affect its whole thickness. These properties have been explored for design of FET biosensing platform [27] as well as electrochemical biosensors [28]. Additionally, it was demonstrated that MoS$_2$ has a high affinity towards ssDNA oligonucleotides and is able to spontaneously adsorb them via van der Waals interactions between nucleobases and the basal plane of MoS$_2$ nano-sheets [29]. This makes MoS$_2$ nano-sheets suitable for capacitive biosensor.

Herein, we report the design of an aptasensor platform in a two-steps process (Scheme 1) by assembling the MoS$_2$ and ssDNA aptamer to sensitively detect acetamiprid. To assure a high stability of DNA on the surface covalent attachment was performed using lipoic acid (LA) self-assembled to MoS$_2$ surface. The sensing platform uses EIS-derived electrochemical capacitance spectroscopy to transduce the recognition event. The novelty in this approach is to perform the electrical measurements without any confined redox probe attached to the surface or in solution as the MoS$_2$ nanosheets exhibit an inherent pseudocapacitance behavior. Upon interaction with acetamiprid, the aptamer will desorb from MoS$_2$ nanosheets to form aptamer/target affinity complex. The aptamer desorption facilitates ionic diffusion of the electrolytes resulting in a signal ON in ECS. The aptasensing electrode allows detecting low levels of the target pesticide and demonstrates high selectivity for the target in presence of competing pesticides. Furthermore, the aptasensor was successfully applied to detect the pesticides in tomatoes purchased form a local market.

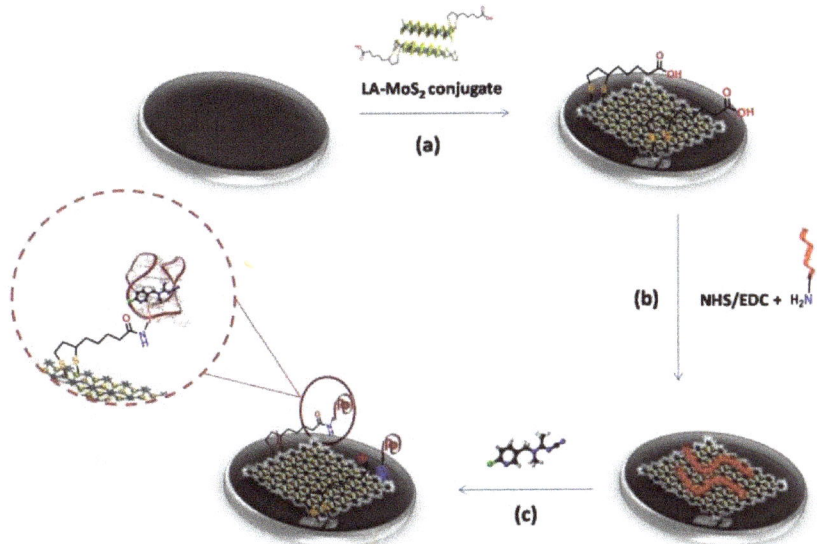

Scheme 1. Building up strategy for the design of acetamiprid aptasensor: (**a**) LA-MoS$_2$ electrodeposition, (**b**) covalent aptamer immobilization, and (**c**) acetamiprid detection.

2. Materials and Methods

2.1. Reagents

(NH$_4$)$_6$Mo$_7$O$_{24}$·4H$_2$O, thiourea, lipoic acid, chloride potassium, lithium perchlorate, N-(3-dimethylaminopropyl)-N-ethylcarbodimide hydrochloride (EDC), acetamiprid, copper (II) sulfate, chlortoluron; were purchased from Sigma-Aldrich. The oligonucleotide was synthesized with an amine modification at the 5′end position and purchased from Sigma Aldrich (Germany). The sequence was originally published by He et al. [30]: 5′-H$_2$N(CH$_2$)$_6$-TGTAATTTGTCTGCAGCGGTTCTTGATCGCTGACACCATATTATGAAGA-3′.

Phosphate-buffered saline (PBS) solutions 0.01 M (pH = 7.4) were prepared by dissolving one tablet in 200 mL of deionized water then filtered using a 0.22 µm membrane filter and stored at 4 °C until use. All chemicals used in this work were of analytical grade and directly used without additional purification. All solutions were prepared with Milli-Q water (18 MΩ cm^{-1}) from a Millipore system. For the real sample assay, tomatoes were purchased from a local market (Tunisia).

2.2. Nanomaterial Preparation

2.2.1. Preparation of MoS$_2$ Nanosheets

The preparation of ultrathin MoS$_2$ nanosheets was achieved by a one-step hydrothermal method following the described procedure [31]. Briefly, 2.28 g (66.3 g/L) of thiourea and 1.24 g (34.4 g/L) of hexaammoniumheptamolybdate tetrahydrate (NH$_4$)$_6$Mo$_7$O$_{24}$·4H$_2$O were dissolved in 36 mL of deionized water to form a homogeneous solution after stirring for 30 min. Then a tightly sealed 50 mL Teflon-lined stainless steel autoclave was filled with the obtained solution and was heated at 220 °C for 24 h. The product was cooled down to room temperature (RT), black precipitates were then collected after centrifugation and washed with distilled water and absolute ethanol for several times. Finally, the obtained MoS$_2$ nanosheets were dried in vacuum at 60 °C for 24 h and characterized with Raman and EDX.

2.2.2. Preparation of LA-MoS$_2$ Conjugate

MoS$_2$ nano-sheets modified with lipoic acid were obtained following the optimized method from literature [32]. Briefly, 32 mg (1.06 g/L) of lipoic acid was dissolved in 30 mL deionized water after fixing the pH at 6.5. Then, 50 mg (1.66 g/L) of MoS$_2$ nano-sheets was added to the resulting solution. The mixture was tip-sonicated (probe tip diameter: 13 mm, VCX 750, Sonics & Materials) for 3 h at 300 W and finally after filtration, LA-MoS$_2$ was obtained (Scheme 2).

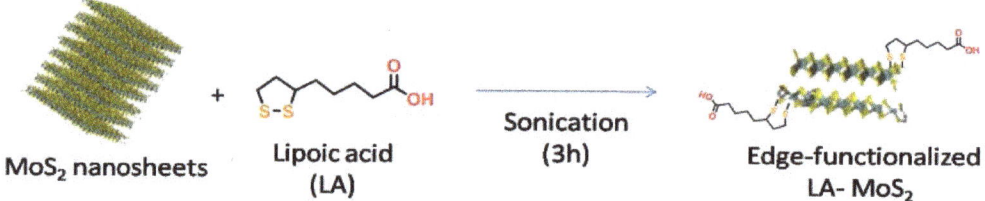

Scheme 2. Schematic description of preparation of LA–MoS$_2$ conjugate.

2.3. Modification of SPCE with MoS$_2$

The screen-printed carbon electrode surface (SPCE) was modified with LA-MoS$_2$ nanosheets by electrochemical deposition. The SPCE was covered with 50 µL of aqueous solution prepared with 5 mg/mL of LA-MoS$_2$ nanosheets dispersed under sonication in 0.5 M LiClO$_4$ solution. Then the potential was swept from 0 to −0.95 V vs. Ag/AgCl at scan rate of 50 mV s^{-1} for 10 cycles. After electrodeposition, the modified electrode was rinsed several times with deionized water and dried under a gentle flux of N$_2$.

2.4. Formation of Acetamiprid Aptasensor

The aptamer functionalized with an amino group in 5′-position was covalently attached to the terminal carboxylic acid present on LA-MoS$_2$/SPCE through an amide bond by incubating the electrode in a solution containing 5 µM of the aptamer in presence of

10 mM of the coupling agent EDC/NHS for 30 min at 35 °C following the optimized method [33]. The electrode surface was thoroughly washed with 0.01 M PBS to remove non covalently attached aptamer. Finally, the biosensor was stored overnight in PBS solution at 4 °C for stabilization.

2.5. Aptamer Binding to Acetamiprid

Aptamers are known to be very stable at RT [34]. Therefore, the aptasensing platform was incubated in 50 µL of acetamiprid solution with different concentrations ranging from 50 to 450 fM for 30 min which is enough time scale to achieve aptamer binding with the target [34]. The electrode was then washed with buffer solution before performing measurements to remove non-attached molecules.

2.6. Electrochemical Measurements

All electrochemical experiments were performed in phosphate buffer saline solutions (PBS, pH = 7.4) using PC-controlled MetrohmAutolab PGSTAT 302n electrochemical workstations with Nova software (v 1.10) to design the experiments and data collection. Screen printed carbon electrodes (SPCE) from Orion High-Tech (Madrid, Spain) were used with a conventional three-electrode configuration a 4-mm diameter-working electrode, a carbon counter electrode and an Ag/AgCl reference electrode.

The measurements were carried out in triplicate by dropping 50 µL of PBS solution onto the SPCE working surface. To characterize the stepwise modification of the surface, 50 µL of 5 mM solution of $[Fe(CN)_6]^{3/4-}$ prepared in PBS solution was dropped on the electrode surface. The AC frequencies for impedance experiments are ranged from 100 KHz to 0.1 Hz with an applied potential of 0.1 V and DC potential of 10 mV. For all the sensing experiments, capacitance curves were measured at a potential of −0.4 V (half-wave potential of −0.4 V (Mo^{4+}/Mo^{3+}) reduction [18] in PBS solution without a redox marker.

The impedance complex $Z^*(\omega)$ was converted into capacitance function $C^*(\omega)$ through the physical equation $Z^*(\omega) = 1/j\omega C^*(\omega)$ in which ω is the angular frequency. The resulting FRA data were processed and treated to obtain the real and imaginary capacitance components respectively from $C'' = \varphi Z'$ and $C' = \varphi Z''$ where $\varphi = (\omega |Z|^2)^{-1}$ and $|Z|$ is the modulus of Z^* [22].

2.7. Methods

Scanning electron microscope (SEM) micrographs and energy dispersive X-ray (EDX) spectra were recorded using a FEI Quanta 200 Environmental SEM. Raman spectra were obtained at room temperature by a Raman Spectrophotometer Horiba Jobin-Yvon equipped with a liquid nitrogen-cooled CCD detector. FT-IR characterizations were obtained using a Bruker Vertex FT-IR spectrometer (Bruker, Germany) equipped with a Mercury cadmium-telluride (MCT) detector and an attenuated total reflectance (ATR) germanium crystal.

3. Results and Discussions

3.1. LA-MoS$_2$ Synthesis and Structural Characterization

MoS$_2$ was obtained by hydrothermal synthesis and was characterized by Raman spectroscopy to confirm the nano-sheets formation. The spectrum of MoS$_2$ shows two characteristic peaks, the out-of-plane vibration of sulfur atoms (A_{1g}) at 405.6 cm^{-1} and a peak characterizing the in-plane vibration of molybdenum and sulfur atoms (E_{2g}^1) located at 382.5 cm^{-1} (Figure 1a). The pic-to-pic difference of 23.1 cm^{-1} is consistent with obtaining one monolayer of MoS$_2$ [35]. To introduce functional group on the surface of MoS$_2$, the nano-sheets were treated with lipoic acid, which forms strong interaction between molybdenum and thiol group. This will provide functional acid group on the surface of MoS$_2$ for further aptamer immobilization.

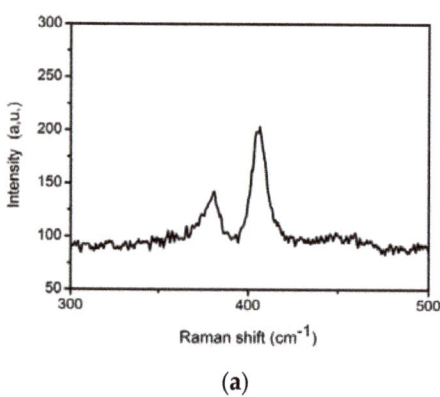 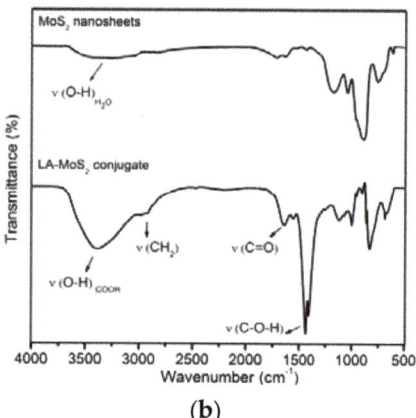

Figure 1. Spectroscopic characterization of MoS_2 and LA–MoS_2: (**a**) Spectrum of MoS_2 after hydrothermal synthesis; (**b**) FTIR spectra of MoS_2 nanosheets and LA–MoS_2 conjugate.

To confirm the presence of lipoic acid moieties on the MoS_2 surface, FT-IR analysis was performed. The spectra displayed in Figure 1b show the characteristic bands attributed to LA such as the band located at 3500 cm^{-1} corresponding to O-H vibrations of carboxylic group, the band at 2925 and 2850 cm^{-1} corresponding to CH_2 and CH stretching vibration and the band 1670 cm^{-1} and 1434 cm^{-1} corresponding to C=O and C-O vibrations, respectively of carbonyl group [36]. All above mentioned bands are absent in the control spectrum of MoS_2. Furthermore, the broad band located at 3392 cm^{-1} corresponds to O-H stretching vibration of residual solvent in the case of unmodified MoS_2 nano-sheets [37].

3.2. LA-MoS_2/SPCE Surface Modification and Characterization

To build the aptasensing platform, the SPCE surface was modified with LA-MoS_2 nano-sheets using an electrochemical deposition method. Indeed, LA-MoS_2 was deposited using a continuous cyclic voltammetry sweep (n = 10) from the corresponding aqueous dispersion, which is an excellent general approach that allows modifying carbon surface by 2D nanomaterials through hydrophobic interaction [38]. This approach has several advantages compared to the conventional deposition methods that begin with precursors molecules. The obtained coating conserved the main properties of the deposited nanomaterials and it can be performed in aqueous solutions under mild condition, at moderate potential, and at RT. Using an electrical field, the chemical environment such as pH changes around the electrode surface due to oxidation or reduction of water. This change results on a decrease of the inter-particle repulsive forces by suppression of the nanomaterial net surface charge, which stabilizes the nanomaterial dispersion and causing its aggregation and irreversible deposition on the electrode surface [39].

The surface morphology before and after SPCE modification was probed using scanning electron microscopy (SEM) to analyze the morphology and energy-dispersive X-ray EDX (EDX) to check the surface composition. SEMs images of bare and LA-MoS_2-modified SPCE are presented respectively on Figure 2a,b. The electrodeposition of LA–MoS_2 led to the formation of multilayer of MoS_2 nano-sheets with aggregates (Figure 2b). The vertical orientation of the MoS_2 nano-sheets could be explained by the interaction between different MoS_2 islands during electrodeposition. The EDX spectrum confirmed the presence of molybdenum and sulfur (Figure 2c). It also showed the Cl and O provided from ClO_4^- used in electrochemical deposition process and remaining in the MoS_2 modified SPCE.

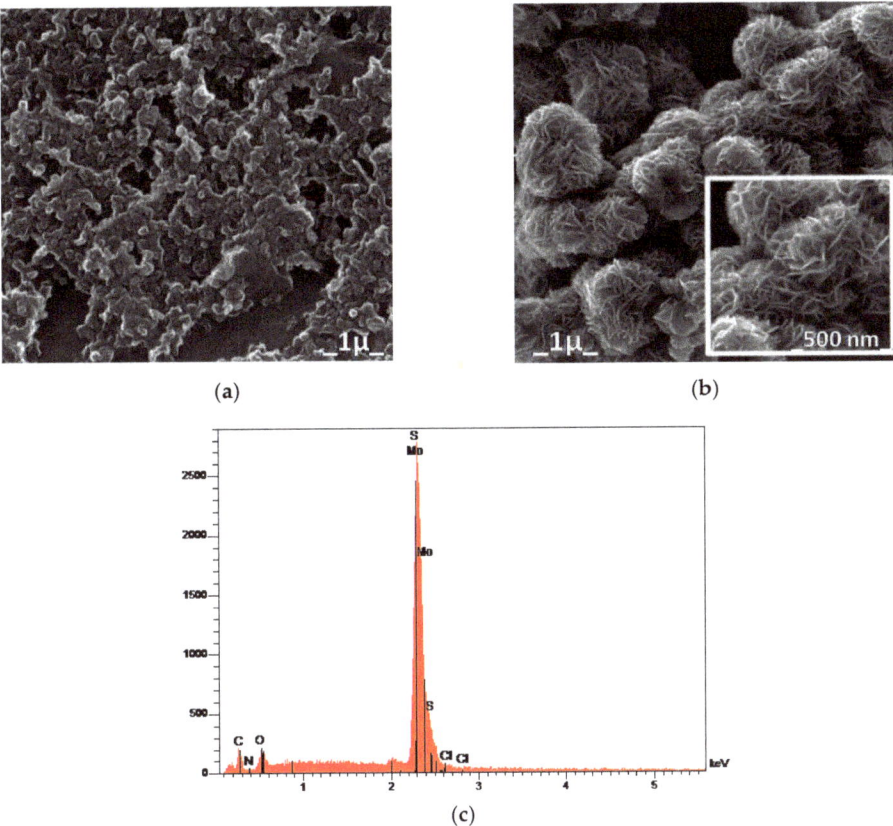

Figure 2. SEM images of: (**a**) bare screen-printed carbon electrode surface (SPCE); (**b**) SPCE modified LA–MoS$_2$ Nano-sheets; (**c**) EDX analysis of the modified SPCE with MoS$_2$ nano-sheets.

3.3. Biolayer Formation

The LA-MoS$_2$ nano-sheets deposited on the SPCE contains carboxylic acid groups provided by LA that can be covalently attached to the aptamer. In a second step, the aptamer was tethered to the surface through an amide bond established between the terminal acid of the self-assembled lipoic acid on the surface of MoS$_2$ and the amino group of the aptamer using NHS/EDC chemistry as depicted in Scheme 1 step b. The SPCE modifications steps were monitored by cyclic voltammetry (CV) and EIS using [Fe(CN)$_6$]$^{3/4-}$ as a redox probe (Figure 3). The CV of LA–MoS$_2$ showed a decrease in the peak current indicating covalent attachment of negatively charged aptamer on the surface of the electrode, which repelled the negatively charged redox probe (Figure 3a, curve b). On the other hand, electrochemical impedance spectroscopy was used to characterize the modified surface. The variation of semicircle diameter of the Nyquist plot (Figure 3b) is related to the charge transfer resistance (R$_{ct}$) and reflects the status of the electrode surface. Modifying the SPCE/LA-MoS$_2$ with aptamer, led to an increase of the R$_{ct}$ (Figure 3b, curve b). This can be explained by the negatively charged aptamer forming a blocking barrier to the diffusion of the redox probe ions thus confirming the CV results. This is in a good agreement with previous reports [40].

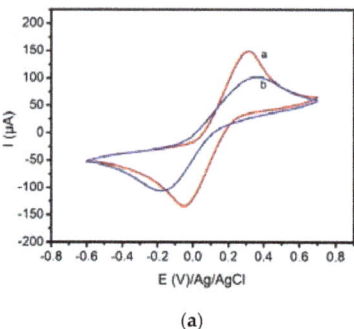
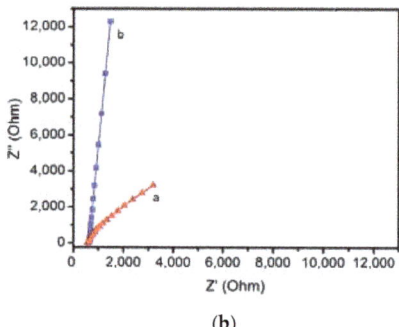

Figure 3. Electrochemical characterization in solution of 5 mM [Fe(CN)$_6$]$^{3/4-}$ and 0.1 MKCl: (**a**) CVs recorded at scan rate 100 mV/s; (**b**) EIS obtained with frequency range: 100 KHz to 0.1 Hz with DC of 10 mV with (curve a) SPCE/LA–MoS$_2$ and (curve b) SPCE/LA–MoS$_2$/APT.

3.4. Analytical Performances

3.4.1. Acetamiprid Detection

The aptasensing platform was incubated in various concentrations of acetamiprid ranging from 50 fM to 450 fM and the ECS was used as transduction method. The capacitance signal, derived from EIS measurements, increased proportionally with the increase acetamiprid concentrations (Figure 4a). The observed behavior can be explained by strong affinity of target analyte to the aptamer, leading to desorption of the immobilized aptamer from MoS$_2$ surface upon formation of the aptamer-target complex. This phenomenon was also observed in the case of DNA hybridization with MoS$_2$ where hybridization reaction led to the desorption of dsDNA from the surface [29]. The loop formation of complex is distant from the surface which facilitates ion diffusion into MoS$_2$ interlayers, where there are more sites available for ion exchange. A linear calibration curve of the average variation of normalized redox capacitance (($C_{0'}-C'/C_{0'}$)*100) with acetamiprid concentration ([ACE]) was plotted (Figure 4b). The latter shows a linear equation regression:

$$\Delta C'/C_0' = 6733 + 0.03[ACE]/(fM) \quad (R^2 = 0.999)$$

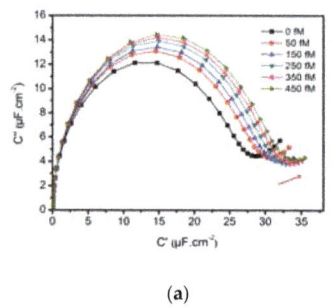
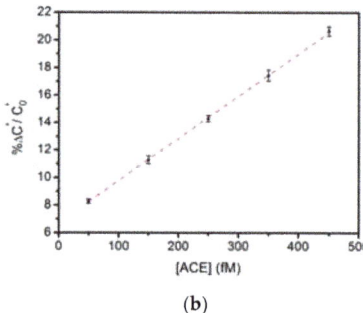

Figure 4. (**a**) Capacitance curves before and after incubation at various acetamiprid concentrations in Phosphate-buffered saline (PBS); (**b**) Calibration curve of the biosensor displaying the relative variation of redox capacitance %$\Delta C'/C_{0'}$ vs. [ACE].

High coefficient of regression in the concentration ranging from 50 to 450 fM is obtained that confirms the linearity of the measurement. Furthermore, the detection limit was calculated to be 14 fM by considering the criteria of signal-to-noise ratio equals three. The reproducibility of the sensor was determined by measuring five different electrodes. The relative standard deviation (RSD) was calculated at 5.4%, which indicates the sensor has good electrode-to-electrode reproducibility thanks to the electrodeposition method which allows high reproducibility of MoS_2 layers in addition of DNA covalent attachment. This biosensor presents a good performance with comparable analytical performance with other systems described in introduction without amplification strategy where signal readout was measured directly after detection. It takes also advantage of the chemical stability of MoS_2 where storage of the MoS_2 and LA-MoS_2 was checked within long time without any variation of properties. In addition aptamer are known to have high stability compared to others biological system [41]. The biosensor is proposed for single use without regeneration.

This detection domain obtained does not include the concentration values of the maximum levels of acetamiprid set by US EPA and EFSA. It is worthy to note that reel samples have to be diluted further before acetamiprid detection, which helps to further decrease the matrix effect. So, taking into consideration this dilution step, it is of high interest to develop aptasensor for acetamiprid detection in diluted extracted samples.

3.4.2. Comparison with Reported Acetamiprid Aptasensors

Regarding the large literature of acetamiprid detection, the analytical performances presented by SPCE/LA-MoS_2 are solely compared with those of published on aptasensing (see Table 1). These studies showed the use of conventional electrochemical techniques such as CV, DPV, and EIS while in this work ECS was used for the first time to achieve sensitive detection of acetamiprid and lead to obtain signal on detection. In terms of LOD the developed analytical device showed the lowest value of 14 fM where the other biosensors are in pM range [15,16,42,43]. The biosensors achieving LOD in the femtomolar range are formed with various nanomaterials including metallic nanoparticles [17,44] which can have negative impact for environment. Moreover, the SPCE/LA-MoS_2 aptasensor presents an easy fabrication process obtained by two steps where patterning through electrodeposition of LA-MoS_2 on SPCE presents an advantage for biosensor conception and chemical attachment of aptamer ensures it to have a longer shelf life. This approach taking advantage of adsorption/desorption process of ssDNA and aptamer complex from MoS_2 surface and the variation of capacitance readout demonstrated excellent balance between high performance, simplicity, and cost-effectiveness compared to others electrochemical biosensors.

Table 1. Analytical performances of various aptasensors.

Platforms	Detection Method	Dynamic Range	LOD	Ref
GCE [1]/AuNPs [2]	CV	0.1 pM–10 nM	0.077 pM	[42]
PtNPs [3] microstrips modified Au IDEs [4]	EIS	10 pM–100 nM	1 pM	[43]
SiNP [5]-streptavidin conjugate modified MB-dsDNA [6]	DPV	500 pM–6.5 μM	1.53 pM	[16]
GCE/rGO-AgNPs [7]/PB-AuNPs [8]	CV	1 pM–1 μM	0.3 pM	[15]
GCE/Au/MWCNT-rGONRe [9]	EIS	50 fM–10 μM	17 fM	[17]
GCE/Ag NPs anchored on nitrogen-doped graphene	EIS	100 fM–5 nM	33 fM	[44]
SPCE/LA-MoS_2	ECS	50 fM–450 fM	14 fM	This work

[1.] GCE:Glassy Carbone Electrode, [2] AuNPs: Gold Nanoparticles, [3] Pt NPs: Platinum Nanoparticles, [4] IDEs: interdigitated electrodes, [5] Si NPs: Silicananoparticles, [6] MB-dsDNA: Methylene Blue-double stranded DNA, [7] rGO-AgNPs: reduced grapheme oxide-Silver Nanoparticles, [8] PB-AuNPs: Prussian blue-gold nanoparticles, [9] MWCNT-rGONRe: Multiwalled carbon nanotubes-reduced graphene oxide nanoribbon.

3.4.3. Selectivity

The selectivity is an important parameter for analytical sensing devices, which characterizes the ability of the aptamer to detect the specific analyte in a sample containing other interfering molecules. The aptasensing platform was challenged by testing it with different interferents such as chlortoluron and copper (II) for copper-based pesticides as only few analytical method could discriminate the nature of remained contaminant residue [45]. Therefore, the biosensor response obtained with acetamiprid alone was compared with those obtained in presence of interferents under the same experimental conditions. The results collected from capacitance curves are presented Figure 5. The response recorded in the presence of aforementioned interferents did not show any increase of the capacitance signal. A blocking effect was observed with chlortoluron, leading to decrease in capacitance while Cu^{2+} induced a minor perturbation of the transduction signal leading also to decrease of ECS response. This study evidences the high selectivity of the aptasensor toward acetamiprid knowing that the interferents have concentrations 10-fold higher than that of the target.

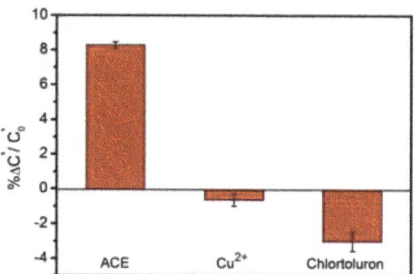

Figure 5. Histograms giving the relative variation of redox capacitance variation after incubation with different interferents.

3.5. Detection of Acetamiprid in Fortified Tomatoe Sample

Tomatoes are one of the foodstuffs that could be affected by the acetamiprid and where the EFSA legislation was fixed the MRL to 0.01 ppm [4]. To perform the detection of acetamiprid on tomatoes, the sample were prepared following the method reported by Kim et al. [46]. Briefly, 5 g of tomatoes samples were extracted with 10 mL of methanol for 30 min, and then centrifuged for 20 min at 10,000 rpm (4 °C) to remove the solids. The supernatant was filtered through a 0.45-micron membrane. Then, aliquots were doped initially with acetamiprid (c_i = 65 fM) that will be confirmed later via the standard addition method. This procedure allows the calibration of analytical devices taking into consideration the matrix effects. The capacitance curves in tomatoes samples solution showed a proportional increase of the capacitance signal with the addition of increasing target concentrations (Figure 6a). The obtained calibration plot (Figure 6b) allowed to determinate c_i by extrapolation to be found equal to 62.1 ± 0.51 fM. In addition, the obtained recovery values for the measurements performed in tomatoes and in PBS are gathered in Table 2. High recovery values are comprised from 95 to 104%, depending on concentration. Thus, the ability of the method to measure small amount of pesticides in food demonstrates the potential of this biosensor of the acetamiprid detection in food.

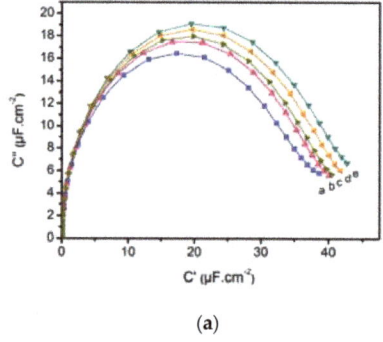
(a)

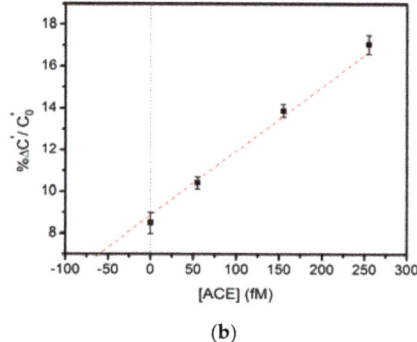
(b)

Figure 6. (a) Capacitance curves before (a: 0 fM) and (b to e) after incubation with various acetamiprid concentrations in fortified tomatoes samples (b: c_i, c: c_i + 55 fM, d: c_i + 155 fM and e: c_i + 255 fM); (b) Calibration curve of the biosensor displaying the relative variation of redox capacitance versus the target concentration.

Table 2. Recovery data for acetamiprid in fortified tomatoes sample.

Sample	[ACE]$_{added}$	[ACE]$_{found}$	Recovery (%)
1 c_i	65.0	62.1 ± 0.51	95.5
2 (c_i + 55.0)	120.0	120.2 ± 0.28	100.1
3 (c_i + 155.0)	220.0	227.0 ± 0.32	103.2
4 (c_i + 255.0)	320.0	334.0 ± 0.45	104.3

4. Conclusions

We report a platform of aptasensing based on MoS$_2$ nanomaterials and the capacitance signal readout for the detection of acetamiprid as a pesticide widely used in agriculture. The device was built on SPCE in two steps to obtain self-assembled MoS$_2$/ssDNA nanostructures. We demonstrated that the high affinity of MoS$_2$ nanosheets for ssDNA and their pseudo redox properties enable the biosensor to achieve rapid detection and high sensitivity. The detection system takes advantage of the adsorption/desorption process provided by the ssDNA and the aptamer complex with the MoS$_2$ surface. In addition, this biosensor has demonstrated high analytical performance with a signal "ON" in the presence of acetamiprid and a detection limit in the fM range, lower than that set by the UPA and EFSA. The study of the detection of spiked acetamiprid in tomatoes showed a measurement comparable with those obtained in a buffer and underlines the ability of this biosensor to measure low levels of pesticides in fresh foodstuffs. These results demonstrated that the methodology developed with these biosensors can be easily adapted to detect other targets of interest. This work paves the way for the development of different highly sensitive and cost-effective aptasensors for food control applications by simply changing the aptamer specific to other pesticides.

Author Contributions: Conceptualization M.H. and N.R.; methodology M.H., N.R., and M.H.; validation, N.R. and H.K.-Y.; formal analysis, M.H.; N.R., and H.K.-Y.; writing—original draft preparation, M.H. and N.R.; writing—review and editing, M.H. and H.K.-Y. Supervision, N.R. and H.K.-Y.; project administration, H.K.-Y.; funding acquisition, H.K.-Y. All authors have read and agreed to the published version of the manuscript.

Funding: This research was funded by MEAE and MESRI PHC grand number 39382RE.

Institutional Review Board Statement: Not Applicable.

Informed Consent Statement: Not applicable.

Data Availability Statement: Data is contained within the article.

Acknowledgments: The authors acknowledge technical support from Instrumental Platform of ICMMO University Paris-Saclay.

Conflicts of Interest: The authors declare no conflict of interest.

References

1. Imamura, T.; Yanagawa, Y.; Nishikawa, K.; Matsumoto, N.; Sakamoto, T. Two cases of acute poisoning with acetamiprid in humans. *Clin. Toxicol.* **2010**, *48*, 851–853. [CrossRef] [PubMed]
2. Kocaman, A.Y.; Topaktaş, M. Genotoxic effects of a particular mixture of acetamiprid and alpha-cypermethrin on chromosome aberration, sister chromatid exchange, and micronucleus formation in human peripheral blood lymphocytes. *Environ. Toxicol.* **2010**, *25*, 157–168. [CrossRef] [PubMed]
3. Qi, Y.; Xiu, F.-R.; Zheng, M.; Li, B. A simple and rapid chemiluminescence aptasensor for acetamiprid in contaminated samples: Sensitivity, selectivity and mechanism. *Biosens. Bioelectron.* **2016**, *83*, 243–249. [CrossRef] [PubMed]
4. U.S. Environmental Protection Agency. Name of Chemical: Acetamiprid Reason for Issuance: Conditional Registration. 2002 Date Issued: March 15 Acetamiprid Proposed Interim Registration Review Decision Case Number 7617 January 2020, Docket Number EPA-HQ-OPP-2012-0329. Available online: https://www.epa.gov/sites/production/files/2020-01/documents/acetamiprid_pid.pdf (accessed on 9 September 2020).
5. Wanatabe, S.; Ito, S.; Kamata, Y.; Omoda, N.; Yamazaki, T.; Munakata, H.; Kaneko, T.; Yuasa, Y. Development of competitive enzyme-linked immunosorbent assays (ELISAs) based on monoclonal antibodies for chloronicotinoid insecticides imidacloprid and acetamiprid. *Anal. Chim. Acta* **2001**, *427*, 211–219. [CrossRef]

6. Zhang, B.; Pan, X.; Venne, L.; Dunnum, S.; McMurry, S.T.; Cobb, G.P.; Anderson, T.A. Development of a method for the determination of 9 currently used cotton pesticides by gas chromatography with electron capture detection. *Talanta* **2008**, *75*, 1055–1060. [CrossRef]
7. Vichapong, J.; Burakham, R.; Srijaranai, S. Alternative Liquid–Liquid Microextraction as Cleanup for Determination of Neonicotinoid Pesticides Prior HPLC Analysis. *Chromatographia* **2016**. [CrossRef]
8. Mateu-Sánchez, M.; Moreno, M.; Arrebola, F.J.; Vidal, J.L.M. Analysis of Acetamiprid in Vegetables Using Gas Chromatography-Tandem Mass Spectrometry. *Anal. Sci.* **2003**, *19*, 701–704. [CrossRef]
9. Zeng, L.; Peng, L.; Wu, D.; Yang, B. Electrochemical Sensors for Food Safety. In *Nutrition in Health and Disease—Our Challenges Now and Forthcoming Time*; IntechOpen: London, UK, 2018. [CrossRef]
10. Li, Z.; Mohamed, M.A.; Vinu Mohan, A.M.; Zhu, Z.; Sharma, V.; Mishra, G.K.; Mishra, R.K. Application of Electrochemical Aptasensors toward Clinical Diagnostics, Food, and Environmental Monitoring: Review. *Sensors (Basel)* **2019**, *19*, 5435. [CrossRef]
11. Ben Aissa, S.; Mars, A.; Catanante, G.; Marty, J.-L.; Raouafi, N. Design of a redox-active surface for ultrasensitive redox capacitive aptasensing of aflatoxin M1 in milk. *Talanta* **2019**, *195*, 525–532. [CrossRef]
12. Zou, X.; Wu, J.; Gu, J.; Shen, L.; Mao, L. Application of Aptamers in Virus Detection and Antiviral Therapy. *Front. Microbiol.* **2019**, *10*, 1462. [CrossRef]
13. Raouafi, A.; Sánchez, A.; Raouafi, N.; Villalonga, R. Electrochemical aptamer-based bioplatform for ultrasensitive detection of prostate specific antigen. *Sens. Actuators B Chem.* **2019**, *297*, 126762. [CrossRef]
14. Sun, D.; Lu, J.; Zhang, L.; Chen, Z. Aptamer-based electrochemical cytosensors for tumor cell detection in cancer diagnosis: A review. *Anal. Chim. Acta* **2019**, *1082*, 1–17. [CrossRef] [PubMed]
15. Shi, X.; Sun, J.; Yao, Y.; Liu, H.; Huang, J.; Guo, Y.; Sun, X. Novel electrochemical aptasensor with dual signal amplification strategy for detection of acetamiprid. *Sci. Total Environ.* **2020**, *705*, 135905. [CrossRef] [PubMed]
16. Taghdisi, S.M.; Danesh, N.M.; Ramezani, M.; Abnous, K. Electrochemical aptamer based assay for the neonicotinoid insecticide acetamiprid based on the use of an unmodified gold electrode. *Microchim. Acta* **2017**, *184*, 499–505. [CrossRef]
17. Fei, A.; Liu, Q.; Huan, J.; Qian, J.; Dong, X.; Qiu, B.; Mao, H.; Wang, K. Label-free impedimetric aptasensor for detection of femtomole level acetamiprid using gold nanoparticles decorated multiwalled carbon nanotube-reduced graphene oxide nanoribbon composites. *Biosens. Bioelectron.* **2015**, *70*, 122–129. [CrossRef]
18. Santos, A. Fundamentals and Applications of Impedimetric and Redox Capacitive Biosensors. *J. Anal. Bioanal. Tech.* **2014**, S7. [CrossRef]
19. Cecchetto, J.; Fernandes, F.C.B.; Lopes, R.; Bueno, P.R. The capacitive sensing of NS1 Flavivirus biomarker. *Biosens. Bioelectron.* **2017**, *87*, 949–956. [CrossRef]
20. Lehr, J.; Weeks, J.R.; Santos, A.; Feliciano, G.T.; Nicholson, M.I.G.; Davis, J.J.; Bueno, P.R. Mapping the ionic fingerprints of molecular monolayers. *Phys. Chem. Chem. Phys.* **2017**, *19*, 15098–15109. [CrossRef]
21. Raouafi, A.; Rabti, A.; Raouafi, N. A printed SWCNT electrode modified with polycatechol and lysozyme for capacitive detection of α-lactalbumin. *Microchim. Acta* **2017**, *184*, 4351–4357. [CrossRef]
22. Fernandes, F.C.B.; Góes, M.S.; Davis, J.J.; Bueno, P.R. Label free redox capacitive biosensing. *Biosens. Bioelectron.* **2013**, *50*, 437–440. [CrossRef]
23. Su, S.; Sun, Q.; Gu, X.; Xu, Y.; Shen, J.; Zhu, D.; Chao, J.; Fan, C.; Wang, L. Two-dimensional nanomaterials for biosensing applications. *TrAC Trends Anal. Chem.* **2019**, *119*, 115610. [CrossRef]
24. Patel, R.; Inamdar, A.; Kim, H.; Im, H. In-Situ hydrothermal synthesis of a MoS2 nanosheet electrode for electrochemical energy storage applications. *J. Korean Phys. Soc.* **2016**, *68*, 1341–1346. [CrossRef]
25. Soon, J.M.; Loh, K. Electrochemical Double-Layer Capacitance of MoS2 Nanowall Films. *Electrochem. Solid-State Lett.* **2007**, *10*, A250. [CrossRef]
26. Chen, X.; Park, Y.J.; Kang, M.; Kang, S.-K.; Koo, J.; Shinde, S.M.; Shin, J.; Jeon, S.; Park, G.; Yan, Y.; et al. CVD-grown monolayer MoS 2 in bioabsorbable electronics and biosensors. *Nat. Commun.* **2018**, *9*, 1690. [CrossRef]
27. Liu, J.; Chen, X.; Wang, Q.; Xiao, M.; Zhong, D.; Sun, W.; Zhang, G.; Zhang, Z. Ultrasensitive Monolayer MoS2 Field-Effect Transistor Based DNA Sensors for Screening of Down Syndrome. *Nano Lett.* **2019**, *19*, 1437–1444. [CrossRef] [PubMed]
28. Dalila R, N.; Md Arshad, M.K.; Gopinath, S.C.B.; Norhaimi, W.M.W.; Fathil, M.F.M. Current and future envision on developing biosensors aided by 2D molybdenum disulfide (MoS2) productions. *Biosens. Bioelectron.* **2019**, *132*, 248–264. [CrossRef] [PubMed]
29. Zhu, C.; Zeng, Z.; Li, H.; Li, F.; Fan, C.; Zhang, H. Single-Layer MoS2-Based Nanoprobes for Homogeneous Detection of Biomolecules. *J. Am. Chem. Soc.* **2013**, *135*, 5998–6001. [CrossRef]
30. He, J.; Liu, Y.; Fan, M.; Liu, X. Isolation and Identification of the DNA Aptamer Target to Acetamiprid. *J. Agric. Food Chem.* **2011**, *59*, 1582–1586. [CrossRef]
31. Zhang, X.H.; Wang, C.; Xue, M.Q.; Lin, B.C.; Ye, X.; Lei, W.N. Hydrothermal synthesis and characterization of ultrathin MoS2 nanosheets. *Chalcogenide Lett.* **2016**, *13*, 27–34. [CrossRef]
32. Jeong, M.; Kim, S.; Ju, S.-Y. Preparation and characterization of a covalent edge-functionalized lipoic acid–MoS2 conjugate. *RSC Adv.* **2016**, *6*, 36248–36255. [CrossRef]
33. Haddaoui, M.; Sola, C.; Raouafi, N.; Korri-Youssoufi, H. E-DNA detection of rpoB gene resistance in Mycobacterium tuberculosis in real samples using Fe3O4/polypyrrole nanocomposite. *Biosens. Bioelectron.* **2019**, *128*, 76–82. [CrossRef] [PubMed]

34. Feier, B.; Băjan, I.; Cristea, C.; Săndulescu, R. Aptamer-based Electrochemical Sensor for the Detection of Ampicillin. In Proceedings of the International Conference on Advancements of Medicine and Health Care through Technology, Cluj-Napoca, Romania, 12–15 October 2016; Vlad, S., Roman, N.M., Eds.; Springer International Publishing: Cham, Switzerland, 2017; pp. 107–110.
35. Li, H.; Lu, G.; Yin, Z.; He, Q.; Li, H.; Zhang, Q.; Zhang, H. Optical Identification of Single- and Few-Layer MoS2 Sheets. *Small* **2012**, *8*, 682–686. [CrossRef] [PubMed]
36. Nie, C.; Zhang, B.; Gao, Y.; Yin, M.; Yi, X.; Zhao, C.; Zhang, Y.; Luo, L.; Wang, S. Thickness-Dependent Enhancement of Electronic Mobility of MoS_2 Transistors via Surface Functionalization. *J. Phys. Chem. C* **2020**, *124*, 16943–16950. [CrossRef]
37. Zhou, K.; Jiang, S.; Bao, C.; Song, L.; Wang, B.; Tang, G.; Hu, Y.; Gui, Z. Preparation of poly(vinyl alcohol) nanocomposites with molybdenum disulfide (MoS_2): Structural characteristics and markedly enhanced properties. *RSC Adv.* **2012**, *2*, 11695. [CrossRef]
38. Kong, N.; Zhang, S.; Liu, J.; Wang, J.; Liu, Z.; Wang, H.; Liu, J.; Yang, W. The influence of 2D nanomaterials on electron transfer across molecular thin films. *Mol. Syst. Des. Eng.* **2019**, *4*, 431–436. [CrossRef]
39. Rastogi, P.K.; Sarkar, S.; Mandler, D. Ionic strength induced electrodeposition of two-dimensional layered MoS_2 nanosheets. *Appl. Mater. Today* **2017**, *8*, 44–53. [CrossRef]
40. Mejri-Omrani, N.; Miodek, A.; Zribi, B.; Marrakchi, M.; Hamdi, M.; Marty, J.-L.; Korri-Youssoufi, H. Direct detection of OTA by impedimetric aptasensor based on modified polypyrrole-dendrimers. *Anal. Chim. Acta* **2016**, *920*, 37–46. [CrossRef]
41. Wang, Y.; Ning, G.; Bi, H.; Wu, Y.; Liu, G.; Zhao, Y. A Novel Ratiometric Electrochemical Assay for Ochratoxin A Coupling Au Nanoparticles Decorated MoS_2 Nanosheets with Aptamer. *Electrochim. Acta* **2018**, *285*, 120–127. [CrossRef]
42. Yao, Y.; Wang, G.X.; Shi, X.J.; Li, J.S.; Yang, F.Z.; Cheng, S.T.; Zhang, H.; Dong, H.W.; Guo, Y.M.; Sun, X.; et al. Ultrasensitive aptamer-based biosensor for acetamiprid using tetrahedral DNA nanostructures. *J. Mater. Sci.* **2020**, *55*, 15975–15987. [CrossRef]
43. Madianos, L.; Tsekenis, G.; Skotadis, E.; Patsiouras, L.; Tsoukalas, D. A highly sensitive impedimetric aptasensor for the selective detection of acetamiprid and atrazine based on microwires formed by platinum nanoparticles. *Biosens. Bioelectron.* **2018**, *101*, 268–274. [CrossRef]
44. Jiang, D.; Du, X.; Liu, Q.; Zhou, L.; Dai, L.; Qian, J.; Wang, K. Silver nanoparticles anchored on nitrogen-doped graphene as a novel electrochemical biosensing platform with enhanced sensitivity for aptamer-based pesticide assay. *Analyst* **2015**, *140*, 6404–6411. [CrossRef] [PubMed]
45. Grimalt, S.; Dehouck, P. Review of Analytical Methods for the Determination of Pesticide Residues in Grapes. *J. Chromatogr. A* **2016**, *1433*, 1–23. [CrossRef] [PubMed]
46. Kim, H.-J.; Shelver, W.L.; Li, Q.X. Monoclonal antibody-based enzyme-linked immunosorbent assay for the insecticide imidacloprid. *Anal. Chim. Acta* **2004**, *509*, 111–118. [CrossRef]

Article

Tonic and Phasic Amperometric Monitoring of Dopamine Using Microelectrode Arrays in Rat Striatum

Martin Lundblad [1,†], David A. Price [1,†], Jason J. Burmeister [1,†], Jorge E. Quintero [1], Peter Huettl [1], Francois Pomerleau [1], Nancy R. Zahniser [2,‡] and Greg A. Gerhardt [1,*]

[1] Department of Neuroscience, Center for Microelectrode Technology and Brain Restoration Center, University of Kentucky Medical Center, MN206, Lexington, KY 40536-0298, USA; martin.lundblad@gmail.com (M.L.); priced0@gmail.com (D.A.P.); jason@jasonburmeisterconsulting.com (J.J.B.); george.quintero@uky.edu (J.E.Q.); peter.huettl@uky.edu (P.H.); francois.pomerleau@uky.edu (F.P.)
[2] Department of Pharmacology, Neuroscience Program, University of Colorado Denver, Aurora, CO 80045, USA; nancy.zahniser@ucdenver.edu
* Correspondence: gregg@uky.edu; Tel.: +1-859-323-4531
† Co-First Authors.
‡ Professor Nancy R. Zahniser died unexpectedly on 5 May 2016.

Received: 7 August 2020; Accepted: 9 September 2020; Published: 16 September 2020

Abstract: Here we report a novel microelectrode array recording approach to measure tonic (resting) and phasic release of dopamine (DA) in DA-rich areas such as the rat striatum and nucleus accumbens. The resulting method is tested in intact central nervous system (CNS) and in animals with extensive loss of the DA pathway using the neurotoxin, 6-hydroxyDA (6-OHDA). The self-referencing amperometric recording method employs Nafion-coated with and without m-phenylenediamine recording sites that through real-time subtraction allow for simultaneous measures of tonic DA levels and transient changes due to depolarization and amphetamine-induced release. The recording method achieves low-level measures of both tonic and phasic DA with decreased recording drift allowing for enhanced sensitivity normally not achieved with electrochemical sensors in vivo.

Keywords: dopamine; sensor; microelectrode array; brain

1. Introduction

Dopamine (DA) serves as a principle neurotransmitter for essential brain pathways that regulate affect, cognition, movement, and reward [1]. Analytical approaches for the detection and quantification of extracellular levels of brain DA have been of the utmost importance for elucidating its role as a modulatory neurotransmitter and for advancing our understanding of how brain DA systems impact ongoing behavior [2,3]. Extracellular DA is present in both synaptic and extrasynaptic pools, which enables differential modulation of pre- and postsynaptic neuronal signaling [4]. Following activity-dependent release, diffusion mediates spillover of synaptic DA into extrasynaptic pools where DA transporters (DATs) are involved in "regulating the sphere of influence and lifetime of released DA beyond a synapse [5]." However, current analytical techniques are unable to measure both tonic and phasic levels of DA simultaneously and, thus, require the use of a complementary approach to investigate the "missing" component of extracellular neurotransmitter dynamics. In turn, there is a need for the effective simultaneous measure and quantitation of synaptic and extra-synaptic DA levels using neurochemical methods that combine real-time monitoring with high spatial-temporal resolution and sensitivity.

Microdialysis sampling coupled with powerful analytical technologies has been the conventional approach for quantifying extracellular neurotransmitter levels in vivo. Once collected, dialysate samples are processed with separation analytics like high-performance liquid chromatography (HPLC), providing measurements with excellent chemical selectivity and sensitivity while affording the opportunity to measure multiple analytes for both acute and chronic in vivo studies [6]. Although high-level chemical selectivity and sensitivity measures have become hallmarks of such strategies, the low-level spatial-temporal resolution of sampling continues to generate curiosity regarding the missed neurochemical dynamics that occur between the large inter-sample intervals (typically 10 min) while spurring controversy as to which extracellular pool of neurotransmitters is being sampled [7,8]. While sampling rates have continued to undergo refinement towards improving temporal resolution (e.g., up to 1 sample per 10 s), second-by-second or sub-second monitoring has not hitherto been achieved [6,9]. Microdialysis probes typically have an active length of 1–4 mm with tip diameters of 0.2–0.3 mm, which places the sampling resolution of this approach at >0.1 mm^3 [6]. Because of its large size, probe-induced tissue damage and neuroinflammation may culminate in secondary effects on neurotransmitter efflux and resting levels, further complicating the interpretation of results [10–13]. Thus, despite the tandem use of powerful separation analytical approaches, the low-level spatial-temporal resolution of microdialysis continues to limit its capabilities for monitoring phasic neurotransmission in vivo.

In vivo electrochemical approaches have emerged as a complementary collection of techniques to microdialysis as they "provide answers that are not presently accessible by microdialysis or any other measurement technique [11]." For example, in vivo electrochemistry has been instrumental to identifying the presence of spontaneous DA release and regulation of DAT-mediated clearance of DA [14–18]. Electrochemical sensors are designed for use in freely moving behavioral recordings where higher resolution is necessary to capture transient DA events. While the chemical selectivity achieved with powerful separation technologies is irrefutable, the spatial-temporal resolution of electrochemical techniques greatly exceeds that of microdialysis probes and, thus, enables sub-second measures with an electrode diameter of 5–200 µm [19]. Importantly, smaller sized electrochemical electrodes produce less tissue responsiveness and damage post-implantation in the brain [20,21]. Although advances in electrochemical approaches have opened up new avenues in the area of in vivo neurotransmitter monitoring including 3D printed carbon electrodes, submicron sized cavity carbon-nanopipette electrodes, sub-millisecond resolution measuring DA over months, and optogenetic stimulation, the most significant shortcoming of these approaches for measures of extracellular DA continues to be the inability to resolve resting neurotransmitter levels [14,22–25].

Here, we describe the development and implementation of ceramic-based microelectrode arrays (MEAs) for intracranial monitoring of tonic and phasic DA neurotransmission by: (1) demonstrating the ability of MEAs to measure DA in vitro and in vivo with constant potential amperometry; (2) discussing in vivo testing through proof-of-concept experiments using normal and denervated striatum in anesthetized rats; and (3) introducing new developments for the measurement of brain DA through a conformal MEA design that enable the real-time, simultaneous monitoring of DA from multiple depths in the rat brain.

2. Materials and Methods

2.1. Reagents

Unless stated otherwise, laboratory chemicals were purchased from Fisher Scientific (Waltham, MA, USA) or Sigma Aldrich (St. Louis, MO, USA).

2.2. Experimental Subjects

Male Sprague–Dawley rats (10–12 weeks old; Harlan Laboratories, Inc.; Indianapolis, IN, USA) were individually housed on a 12 h light/dark cycle with ad libitum access to food and water. Animals were acclimated ≥1 week before any experiment. All procedures involving the use of animals were carried out in accordance with the National Institutes of Health Guide for the Care and Use of

Laboratory Animals and were approved by the Institutional Animal Care and Use Committee of the University of Kentucky (IACUC Identification Number: 2019-3387).

2.3. Dopamine (DA) Lesions

Striatal DA was depleted by unilateral infusion of 6-hydroxydopamine (6-OHDA) into the medial forebrain bundle (MFB), which causes an extensive and non-recoverable lesion to the DA neurons [26,27]. Animals were anesthetized by inhalation of 1.5–3% isoflurane (Isothesia™, Butler; Dublin, OH, USA) and, once stable, were administered Rimadyl® (5 mg/kg, intraperitoneal injection (i.p.)) for pre-operative analgesia. Body temperature was maintained using a recirculating water blanket (Gaymar® Industries, Inc.; Orchard Park, NY, USA). The right MFB was infused with 6-OHDA (3.0 µg/µL) or vehicle—sterile saline containing 0.02% ascorbic acid (AA)—at a flow rate of 1 µL/min through a 26 s gauge GASTIGHT Hamilton syringe (Hamilton Company; Reno, NV). The delivered volumes and stereotaxic coordinates were 2.5 µL at tooth bar (TB): −2.3 mm, anterior-posterior (AP): −4.4 mm, medial-lateral (ML): −1.2 mm, dorsal-ventral (DV): −7.8 mm and 2.0 µL at TB: +3.4 mm, AP: −4.0 mm, ML: −0.8 mm, DV: −8.0 mm. The AP and ML stereotaxic coordinates were taken with respect to Bregma and DV coordinates were determined relative to the brain surface for all stereotaxic surgeries [28]. Following infusion, the syringe was left in place for 2.5 min before being slowly retracted.

2.4. Microelectrode Array (MEA) Preparation

The photolithographic fabrication of ceramic MEAs has previously been described in detail [27,29]. All electrochemical procedures were carried out using the FAST-16 mkII recording system and software (Quanteon, L.L.C.; Nicholasville, KY, USA). The present study used S2 (Side-by-Side 2nd generation, Figure 1A) and Double-Sided-Pair-Row-8 pairs, conformal DSPR8 (Figure 1B) ceramic-based MEAs which were obtained from the Center for Microelectrode Technology cost center (University of Kentucky, Lexington, KY, USA). S2 MEAs, have previously been characterized [25]. These comprised n = 4 platinum recording sites (15 × 333 µm each) geometrically arranged as two side-by-side pairs (30 µm between sites within a pair, 100 µm separation between pairs) [30]. The newly designed, conformal DSPR8 MEAs comprised n = 8 platinum recording sites (50 × 100 µm each) geometrically arranged as four vertically aligned pairs (100 µm between sites within a pair) with differential spacing between pairs to enable simultaneous electrochemical recordings at multiple brain depths along the dorsal-ventral plane (Figure 1B). All MEAs were prescreened to select electrodes that had recording site sensitivities for analytes that were within +/− 10% standard deviation (SD) to achieve the analytical performance necessary for the studies. On the day before electrochemical recordings, all Pt recording sites were coated with the anionic polymer Nafion® (Figure 1A,B), which repels negatively charged interferents such as AA and 3,4-dihydroxyphenylacetic acid (DOPAC), as previously described [29,31,32]. One site within each pair of S2 MEAs (Figure 1A) or the upper recording site of each pair of the DSPR8 MEAs (Figure 1B) was then selectively electroplated with *m*-phenylenediamine (m-PD), a size exclusion barrier that blocks DA and other large molecules from reaching the Pt recording sites [33]. m-PD was selectively electrodeposited onto individual platinum recording sites as previously described [34]. Following coating with Nafion® and differential electroplating with m-PD, MEA tips were soaked in phosphate-buffered saline (PBS) at 25 °C overnight prior to use to remove excess m-PD molecules. The final configuration of S2 (Figure 1A) and DSPR8 (Figure 1B) MEAs consisted of Nafion®-only sites (i.e., DA sites) and Nafion®-coated + m-PD electroplated sites (i.e., sentinel sites).

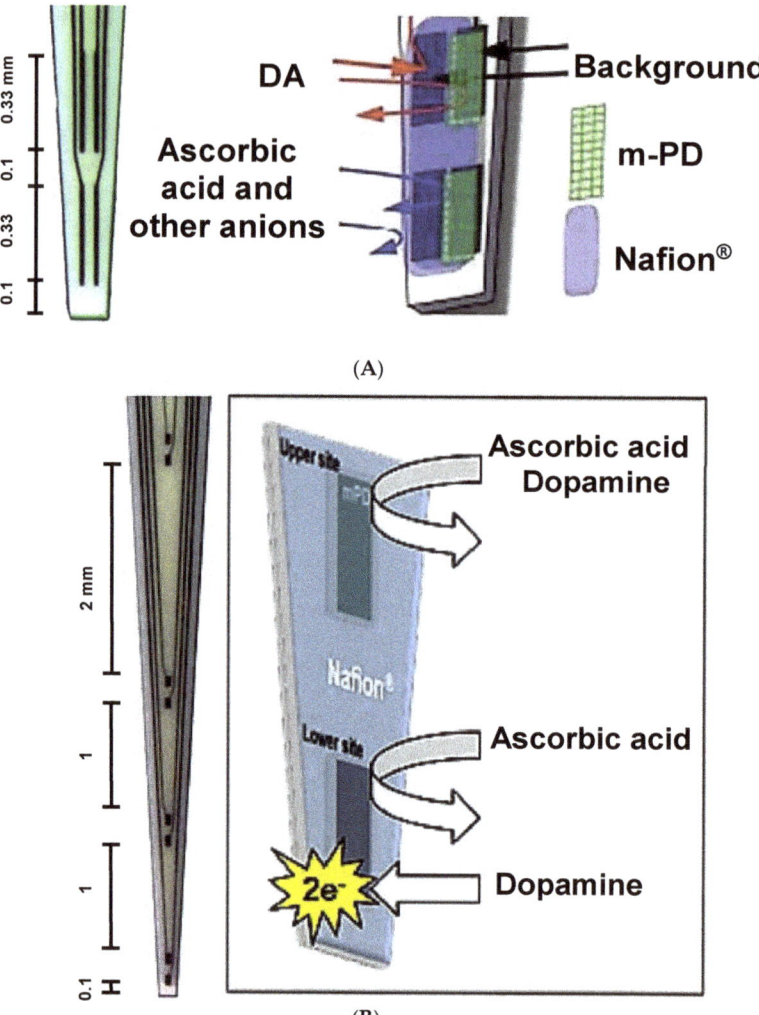

Figure 1. Microelectrode array (MEA) design and function as configured for dopamine (DA) measurement. Photomicrographs and corresponding schematics of Nafion®-coating and selective electrodeposition of *m*-phenylenediamine (m-PD) for a (**A**) 4-channel S2 MEA (333 × 15 micron pairs) and (**B**) 8-channel DSPR8 MEA with reaction schemes are shown. The distance between recording pairs is also shown for both MEAs.

2.5. In Vitro Microelectrode Array Electrochemistry

On the day of in vivo electrochemical recordings, MEAs were individually calibrated in 40 mL of stirred 25 °C PBS (0.05 M; pH 7.4) as previously described [30]. Constant potential amperometry was used for all electrochemical experiments. A +0.35 V potential vs. a RE-5B Ag/AgCl reference (Bioanalytical Systems; West Lafayette, IN, USA) was applied at a frequency of 1 Hz (S2) or 4 Hz (DSPR8); 1 Hz and 4 Hz data were comparable. After establishing a stable baseline, interferent(s) and analyte additions were made to the stirred solution. Unless stated otherwise, AA (250 µM) and three additions of DA (2 µM increments) served as the interferent and analyte, respectively.

2.6. In Vivo Microelectrode Array Electrochemistry

Surgical procedures for in vivo electrochemical studies were carried out as previously described [35,36]. Animals were anesthetized with urethane (1.25 g/kg, i.p.), and body temperature was maintained using a re-circulating water blanket. A Ag/AgCl reference wire was placed in the superficial cortex through a hole drilled caudal to the MEA implantation site. Calibrated S2 MEAs were first targeted to the primary motor cortex using the following coordinates: AP: +1.0 mm, ML: ±2.5 mm, DV: −2.0 mm. After establishing a baseline signal in cortex, S2 MEAs were lowered ventrally with the aid of a microdrive to the dorsal striatum (dSTR) to a new DV coordinate of −3.5 mm. DSPR8 MEAs were implanted using the following coordinates: AP: +1.0 mm, ML: −2.5 mm, DV: −6.3 mm. In all cases, the tip of the MEA device was lowered to the indicated DV coordinate with respect to the brain surface. Once implanted, saline-soaked cotton balls or gauze were placed around the MEA shank to keep the brain surface moist. For some studies, micropipettes were affixed to S2 MEAs (aligned in the center between upper and lower recording pairs) for the local delivery of isotonic KCl (120 mM KCl, 20 mM NaCl, 2.5 mM $CaCl_2$; pH 7.4) as previously described [34]. At the end of in vivo electrochemical recordings, each animal was sacrificed by decapitation.

2.7. Determination of Striatal DA Tissue Content

The tissue content of DA in the dorsal striatum was determined from normal and 6-OHDA lesioned animals as previously described [37].

2.8. Data and Statistical Analyses

All data are expressed as the mean ± standard error of the mean (SEM). Means were determined by summing the individual samples then dividing by the number of samples. Standard deviation was calculated by taking the square root of the variance. Variance was calculated by averaging the squared differences from the mean. SEM was calculated by dividing the standard deviation of the samples by the square root of the sample size. Selectivity, linearity (R^2), sensitivity (i.e., DA slope) and limit of detection (LOD) were calculated for all recording sites as previously described [29]. In the case of S2 MEAs, self-referencing was used to identify and eliminate interfering signals and background charging current from the analyte response as previously described [32]. Thus, signals from sentinel sites were subtracted from those of DA sites of S2 MEAs to yield one self-referenced signal per pair of platinum recording sites (i.e., the self-referenced signal). Stable baseline recordings were first obtained from the primary motor cortex (i.e., an area of low DA innervation) in each animal before recording from the dorsal striatum (i.e., area of high DA innervation) [38,39]. The average current of the self-referenced cortical signal 1 min prior to relocation to the dSTR was subtracted from each point of the self-referenced dSTR signal of the same animal to determine the resting DA current.

3. Results

3.1. In Vitro Characterization of MEAs for Measuring DA

The recording properties of 4-channel S2 (Figure 1A) and 8-channel DSPR8 (Figure 1B) MEAs configured for the measurement of DA were evaluated through in vitro calibrations in PBS (Table 1). Constant potential amperometry ($E_{applied}$ = +0.35 V) was used to measure DA at individual recording sites over a DA concentration range of 0–6 µM (2 µM increments). Oxidation of DA with the MEA recording sites occurred at any potential greater than +0.2 V vs. Ag/Ag+ on the MEAs [40]. The applied potential of +0.35 V was chosen because it achieved a robust DA response while limiting oxidation of species that could be present like H_2O_2 or NO that oxidize at higher potentials. Nafion®-coated DA recording sites on both S2 and DSPR8 MEAs had excellent linearity for DA detection while effectively maintaining high selectivity ratios for DA over 250 µM challenges of AA (Table 1) or DOPAC (data not shown). As shown in Table 1, both MEA types demonstrated high-sensitivity measurements of DA (range, in pA/µM: S2, −5.9 to −121.3; DSPR8, −3.5 to −107.2), nanomolar limits of detection recorded without self-referencing subtraction (range, in nM: S2, 2.7 to 430; DSPR8, 1.9 to 100.0) and showed similar sensitivity per unit area measurements of DA (range, in pA/µM^{-1}/mm^{-2}: S2, −1181 to −24,277; DSPR8, −696 to −21,437). Importantly, sentinel recording sites (i.e., Nafion®-coated + m-PD) did not respond to additions of DA or AA. Sensitivity/area is comparable to previous work [29].

Table 1. Recording properties of microelectrode arrays for the measurement of DA.

Electrode Type	n	Sensitivity (pA/µM)	Sensitivity/Area (pA mM^{-1} mm^{-2})	LOD (nM)	R^2	Selectivity
S2	23	−42.0 ± 0.0	−8422 ± 954	62.3 ± 21.3	0.976 ± 0.008	1487 ± 194
DSPR8	20	−31.7 ± 7.1	−6349 ± 1423	27.5 ± 5.7	0.965 ± 0.006	2664 ± 358

3.2. Proof-of-Concept for MEA-Based In Vivo Measures of Phasic and Tonic DA Levels

The high level of spatial-temporal resolution imparted by in vivo electrochemical technologies with microelectrodes has been instrumental to shaping current views of phasic DA neurotransmission in the areas of release and uptake [11]. The ability of S2 MEAs to measure phasic changes in extracellular DA was determined in the dSTR of anesthetized rats with and without a 6-OHDA lesion. The in vivo MEA response was tracked following the local application of DA (100 nL; 200 µM, pH 7.4) in the dSTR (data not shown). In addition to detecting exogenous DA, MEAs successfully captured active DA uptake—indicated by the descending portion of the curve. In keeping with previous observations from our laboratory using carbon fiber microelectrodes, amplitude-matched DA signals were cleared more slowly in the lesioned vs. non-lesioned dSTR [41]. Figure 2 shows the in vivo MEA response to repeated applications of KCl (100 nL; 120 mM, pH 7.4) in the dSTR, which stimulates neurotransmitter release from nerve terminals. Ejection of KCl in the non-lesioned dSTR stimulated endogenous DA release (peak amplitude in µM: first stimulation, 4.3; second stimulation, 3.1). Importantly, KCl-evoked DA release in the lesioned dSTR was significantly reduced for both the first and second applications of KCl (peak amplitude in µM based on the calibration slope calculated in pA/µM: first stimulation, 0.1; second stimulation, 0.05). In addition, while the rat striatum has norepinephrine and serotonin nerve terminals that can be detected by the MEAs, the KCl-evoked signals are dramatically reduced in the lesioned striatum showing that the signal recorded is likely predominantly DA. The MEAs are known to respond to serotonin and norepinephrine in vitro and in vivo, but the levels of these neurotransmitters in vivo are 40–1000 folds lower than the DA based on the tissue levels of the monoamines [42]. Collectively, these data support the use of MEAs for traditional measures of in vivo DA signals.

The inability of in vivo electrochemical techniques to measure tonic DA levels continues to be a major shortcoming and necessitates that a complementary approach (e.g., microdialysis) be used to study resting DA levels [7]. While the combined approach escalates the complexity, cost and commitment to a study, inclusion of one approach and not the other may only be appropriate for certain experimental designs. MEAs have been routinely utilized to quantify both evoked and resting glutamate levels in the CNS [30,32–34,43–46]. Therefore, we employed 4-channel S2 MEAs in initial studies aimed at providing proof-of-concept for measurement of tonic DA levels in the dSTR of anesthetized rats. Preliminary studies determined that self-referencing alone did not always remove the intrinsic background current of S2 MEA recording sites in vivo (data not shown). Therefore, we utilized the small size of the MEA device to record electrochemical signals in brain regions that differ with regard to dopaminergic innervation—DA innervation is low in the primary motor cortex and high in the dSTR—in conjunction with the self-referencing approach to establish a baseline electrochemical signal in the brain microenvironment (Figure 3A). MEAs were first positioned in the cortex, and after establishing a stable baseline signal (typically ≥90 min), S2 MEAs were lowered ventrally to a final recording position in the dSTR (Figure 3A). Initially, the self-referenced baseline signal measured by DA sites appeared to be comparable between the cortex and dSTR (Figure 3B). However, Figure 3B shows that the dSTR baseline signal is actually greater (i.e., flipped) vs. the cortical baseline signal (in vivo baseline current, in pA: cortex, 71.7; dSTR, 73.0). Thus, by first recording in the primary motor cortex, which is void of DA, the electrochemical current in the dSTR and background electrochemical current of the brain could be distinguished. We next used these basic principles to measure tonic (i.e., resting) DA levels in the dSTR of rats with and without a 6-OHDA lesion. Following self-referencing, the cortical baseline signal was subtracted from the dSTR baseline signal to remove the background electrochemical current to enable resting DA levels to be quantified. The S2 MEA approach showed a significant 88% decrease of resting DA levels in the dSTR (in nM: non-lesioned, 19.4 ± 2.1; 6-OHDA, 2.3 ± 1.3 ***; paired t-test, *** $p < 0.001$). HPLC confirmed a 99% loss of DA tissue content in the denervated dSTR (in µg/g of tissue: non-lesioned, 6.3 ± 0.8; 6-OHDA, 0.03 ± 0.002). These low nanomolar levels of resting DA are consistent with studies using no net flux microdialysis [6,47–50] and the recent square wave voltammetry study by Taylor et al. [51].

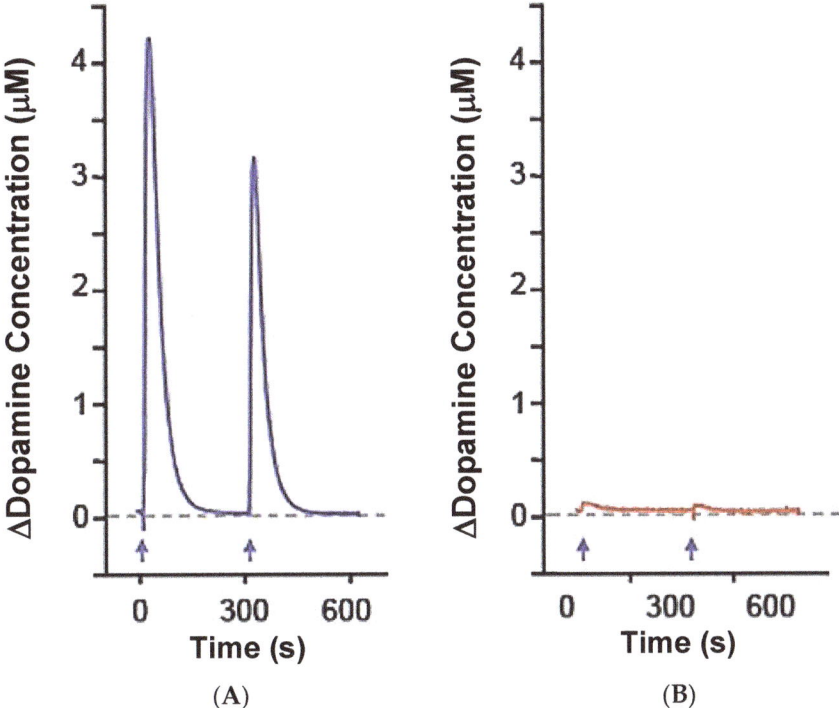

Figure 2. In vivo MEA-based measures of phasic DA neurotransmission. Representative traces of the DA signal measured on DA (solid line) and sentinel (dashed line) recording sites following the local application of KCl (100 nL; 120 mM, pH 7.4) in the dorsal striatum (dSTR) of anesthetized non-lesioned ((**A**)–Blue) or 6-hydroxydopamine (6-OHDA) lesioned ((**B**)—Red) rats.

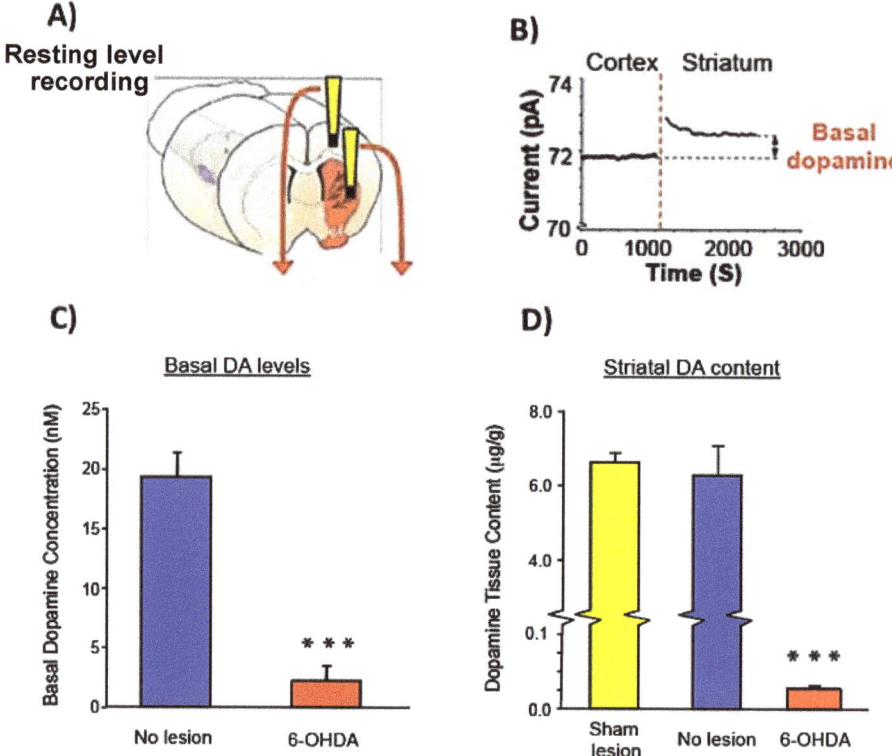

Figure 3. Proof-of-concept for MEA-based measures of resting DA levels in vivo. (**A**) Reconstruction of differential cortical-striatal recording with a 4-channel S2 MEA. S2 MEAs were positioned in primary motor cortex and after establishing a baseline signal were lowered ventrally into the dSTR. (**B**) Baseline current differences measured in the anesthetized rat primary motor cortex and dSTR following self-referencing. Recording shows the cortical-striatal baseline current increase between the primary motor cortex (i.e., low DA innervation) and dSTR (i.e., high DA innervation). The difference between the baseline current measured in the dSTR vs. cortex represents resting DA levels. (**C**) The self-referenced cortical baseline was subtracted from the self-referenced baseline signal in dSTR to calculate differences in resting DA levels between the non-lesioned and 6-OHDA lesioned dSTR. *** $p < 0.001$, paired t-test. Note the low nanomolar detected levels of DA by the method. (**D**) High-performance liquid chromatography (HPLC) analysis of DA tissue content from non-lesioned and 6-OHDA lesioned dSTR. *** $p < 0.001$, paired t-test (n = 5–7).

3.3. Conformal MEAs for In Vivo Measures of DA from Multiple Recording Depths

The proof-of-principle studies outlined above led us to design a novel 8-channel MEA with differential spacing of platinum recording site pairs to enable simultaneous measures of DA at multiple recording depths in the rat brain. The single-sided DSPR8 MEA was designed on earlier studies using S2 MEAs to measure DA. Figure 4A shows the 1–2 mm spacing between recording pairs on the DSPR8 MEA, which allows, in this case, one pair of sites to record in the primary motor cortex while the other three more ventral pairs (≥2 mm from the cortical pair) are positioned to record at multiple depths in the dSTR (1 mm spacing between each pair in the dSTR). Once the tip of the DSPR8 MEA was lowered to the final DV coordinate, the electrochemical signals were allowed to reach a stable baseline (typically ≥90 min). Figure 4B–D show the self-referenced DA signal at each recording depth before and after systemic administration of the psychostimulant d-amphetamine (2 mg/kg, i.p.).

The onset of d-amphetamine changes in DA signaling was quite rapid. In vitro testing showed MEAs are insensitive to d-amphetamine.

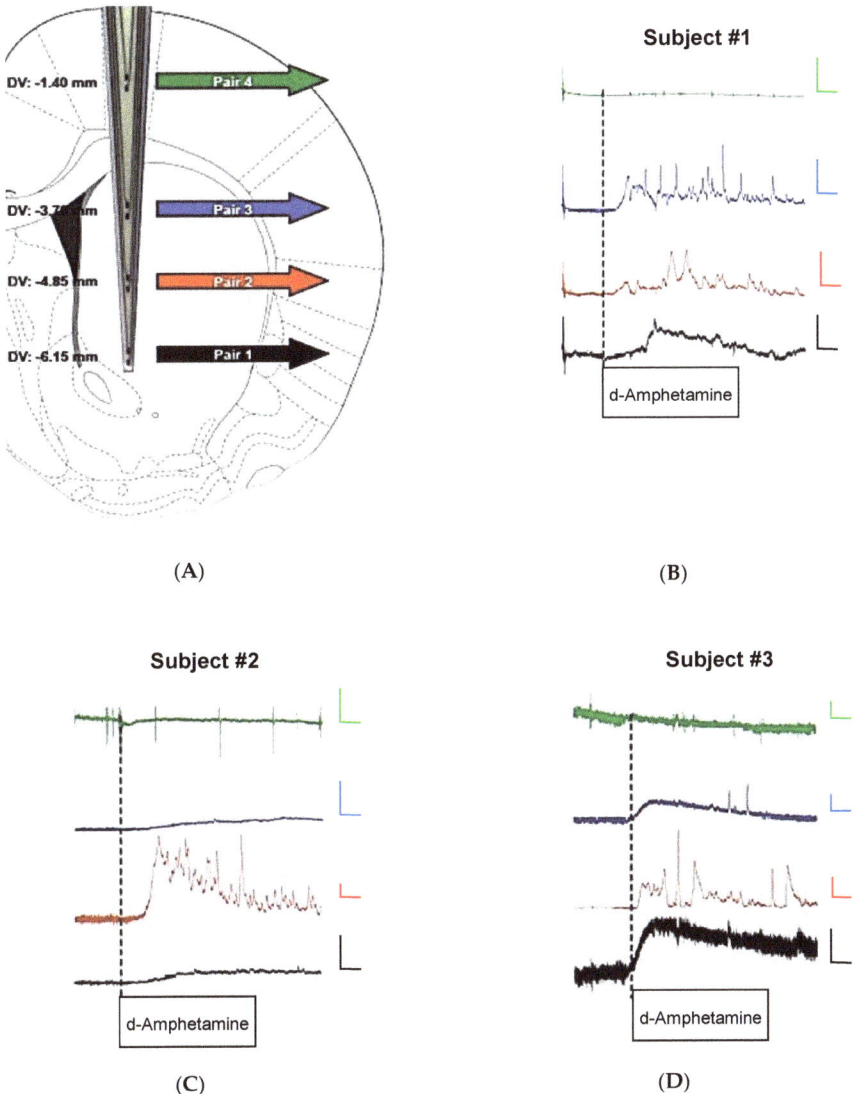

Figure 4. Conformal MEA design for simultaneous measures of in vivo DA dynamics at multiple recording depths. (**A**) Depth profile illustrating location of each MEA pair; (**B**–**D**) concentration versus time plots for subjects 1–3 before and after administration of d-amphetamine (2 mg/kg, i.p.). Time axis bars are 10 min and concentration axis bars are 100 µM.

The effects of the psychostimulant, d-amphetamine, on DA release have been widely characterized and are a known way to measure non-calcium dependent release of DA [52]. Therefore, we assessed the effect of systemic administration of d-amphetamine (2 mg/kg, i.p.) on resting levels of DA in the anesthetized dSTR of rats. d-Amphetamine caused a rapid extracellular increase in DA. Interestingly, the elevated levels of resting DA were accompanied by DA transients seen more closely in Figure 5, which are likely related to the complex actions of d-amphetamine on DA release. The number of spikes and the amplitudes varied in number and size (range: ~5 nM to 800 nM). Of particular note is that this pilot study shows the variable response of DA nerve terminals to the d-amphetamine in subregions of the rat striatum and nucleus accumbens (ventral striatum) of the same animal.

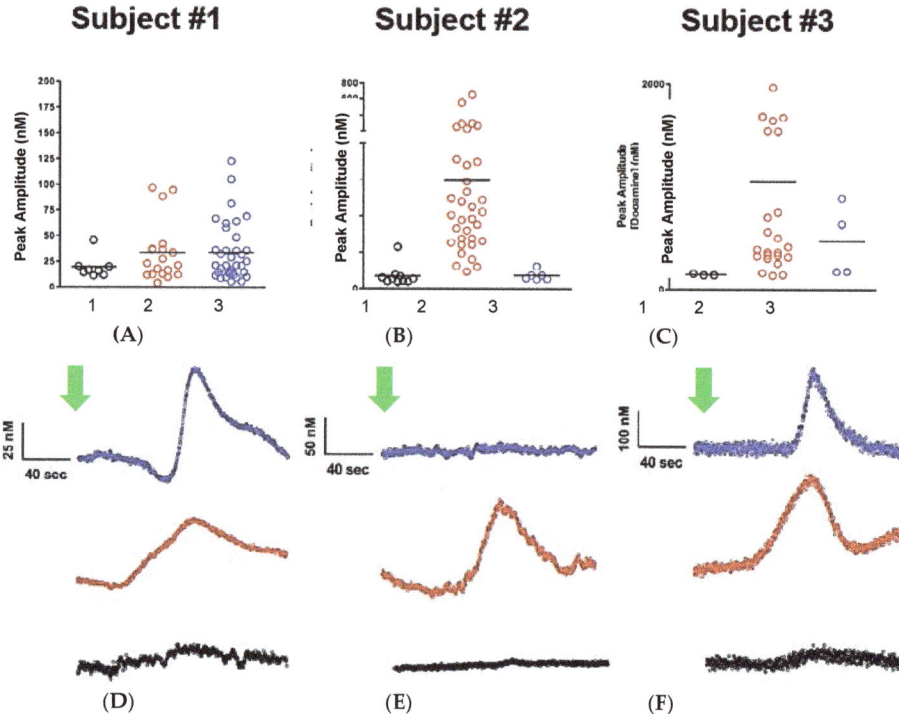

Figure 5. Upper plots (**A**–**C**) show the self-referenced dopamine peak amplitude at the 3 depths for subjects 1–3; bar is the average amplitude of the spikes in each region. Lower traces (**D**–**F**) are self-referenced DA concentration versus time plots for the three test subjects following administration of d-amphetamine. Green arrows indicate the time that the d-amphetamine was administered. Note the high signal-to-noise of these signals that are not filtered or averaged. Typical detection limits of these signals, based on a signal-to-noise of 3, typically ranged from 0.3 to 1 nanomolar, rivaling the recent report of Taylor et al., 2019 [51].

4. Discussion

In vivo monitoring of neurotransmitters in the CNS during the last decade with ceramic-based MEAs has primarily focused on the measurement and quantification of non-electroactive chemical species (e.g., glutamate) through enzyme-mediated conversion to an electroactive reporter molecule (i.e., H_2O_2) [30,34,43–45]. Undoubtedly, the success of these approaches for the determination of phasic and tonic neurotransmitter levels, can be directly attributed to the multi-site configuration of the MEAs [29,32]. More recently, Nafion®-coated MEAs with m-PD electroplated on select platinum recording sites were used for the measurement of brain nitric oxide [53]. Here, S2 MEAs were successfully utilized to measure phasic and tonic DA levels in the anesthetized rat dSTR—extending the abilities of MEA technology for in vivo monitoring. The confirmation of a hypodopaminergic state in the 6-OHDA lesion model—attenuated DA uptake, impaired evoked release of DA and depleted resting DA levels—observed with the S2 MEAs provided a sound platform and proof-of-concept for further exploring the utility of this approach.

In addition to repelling negatively charged molecules (e.g., AA), thus reducing high-level background currents in vivo, Nafion® concentrates positively charged molecules (e.g., DA) near the electrode surface [28,31,32]. During the last decade, the design and fabrication of MEAs has continued to evolve alongside steps towards understanding the unique performance abilities of the device [54]. When compared with earlier MEA designs, the enhanced performance of the MEA types used here to measure DA can likely be attributed to improved precision and reproducibility of the Pt recording sites due to the MEA fabrication process [29]. Subtle differences between individual recording site performances within a single MEA device, as well as between MEA devices, are not unexpected since individual pads retain unique physical properties post-fabrication. In this regard, we have recently shown that the nanostructured surface topography of individual pads contributes to an electroactive area that is greater than the geometric area [41]. However, dissimilarities regarding the thickness of Nafion® following dip-coating, which may produce differential diffusion layers, may also contribute to small differences in the responsiveness of individual recording sites. In addition to confirming the utility of Nafion® for repelling major CNS interferents, our findings here further support the utility of m-PD for removing larger organic molecules (e.g., AA), including the analyte of interest (i.e., DA), for the incorporation of self-referencing approaches [33].

Both Nafion and m-PD were employed on the sentinel recording sites to achieve the self-referencing configuration to selectively measure resting DA levels This dual-selective layer makes the Pt recording sites of the MEAs essentially unresponsive to electroactive species that are anionic and larger than H_2O_2. When the sentinel signal is subtracted from the Nafion-only site a 'pure' catecholamine signal is achieved. In addition, m-PD can be precisely coated onto recording sites even if they are in close proximity (within microns) to other sites. The resulting electrode pairs have excellent selectivity over anionic species such as DOPAC and ascorbate. In addition, they display nanomolar detection limits for DA which are adequate to measure resting levels. However, the subtraction approach employed in this study achieved improved baseline stability for long recordings in vivo (>2 h) and had an improved apparent limit of detection (LOD) that far exceeds standard amperometric recordings with the same MEAs (Table 1). This is similar to self-referenced in vivo glutamate measures where noise that is present on both sites is removed [55,56]. Our estimated LOD for the S2 and DSRP8 MEAs, employed using the self-referencing subtraction approach, ranged from 0.3 to 1 nanomolar, which rivals the recent report of DA detection by Taylor et al. using square-wave voltammetry [51]. This is seen from the recordings of d-amphetamine-induced DA transients seen in Figure 5, which required no filtering or signal averaging. The improved stability and enhanced LOD of the recording are attributed to the "real time" subtraction of the non-Faradaic background current of the Pt recording pairs that achieves the measures of resting DA. The improved baseline stability and the enhanced LOD exceed the capabilities of other amperometric recording methods, warranting further investigation and use.

Amperometric recording methods have always been limited due to the inherent problems of the current signal being composed of both non-Faradaic, background current and the Faradaic response to analytes [57]. As such, techniques such as differential pulse voltammetry and chronoamperometry were developed to minimize the analytical shortcomings of the unpredictability of an electrode's "double layer" [57]. The present study employed a novel approach to subtract off the apparent "non-Faradaic" signals of the recording pads by a real-time subtraction approach that measures the differential response of two nearly identical recording pads. This approach cannot be readily achieved by hand-fabricated microelectrodes where the active surface area of the electrode cannot always be predicted from the geometric area of the surface. By contrast, the MEAs are a microfabricated technology that achieves precision by methods used in the microelectronic industry to build microchips. The manufacturing process achieves a high level of precision and the MEAs are sorted for their analytical response to achieve precision so that in essence the Pt recording sites are nearly identical. Thus, in contrast to many microelectrode technologies, the active recording surfaces are highly reproducible and the active recording area of the electrode is proportional to the surface area. We stress that the differential recording method described in this paper will only work with MEAs that have a very high level of recording site precision. In this study, conformal MEAs were used to compare real-time DA levels from multiple recording depths in vivo within the same animal following systemic administration of d-amphetamine. D-amphetamine has been extensively studied and is known to produce a non-calcium-dependent release of DA through its interaction with the dopamine transporter (DAT) located on the DA presynaptic terminal [52]. Importantly, DA is released through the reverse transport of DAT [21]. Our study demonstrated such release in multiple areas of the rat striatum, yielding signals that contained both slower DA release and evidence of transient DA signals of varying amplitudes and time courses in some brain regions. Galli and coworkers using electrophysiological methods to study DAT have shown that d-amphetamine can change the probability of d-amphetamine-induced DA release through the DAT by a shuttle carrier vs. pore like model [58]. We attribute the more transient spikes of DA release seen in some to the rat brain areas to be due to enhanced pore function of the DAT. This is an exciting hypothesis that needs to be further explored.

5. Conclusions

The present study has demonstrated that a differential recording method can be employed in DA-rich areas of the rat striatum and nucleus accumbens to reliably measure tonic (resting) and phasic DA release with subsecond temporal resolution and a refined spatial resolution dictated by the size of the MEA Pt recording sites. Proof of concept studies show that the technique can reliably measure changes in resting levels and transient changes in DA with improved baseline stability and a greatly enhanced apparent LOD. The use of the DSRP8 MEAs demonstrates that multisite recordings can be simultaneously made in brain subregions within the same animal. The enhanced stability of the recording method and the improved LOD of the method is worthy of further exploration.

Author Contributions: Conceptualization, G.A.G., N.R.Z. and M.L.; methodology, F.P. and J.E.Q.; software, J.J.B.; validation, M.L., D.A.P. and J.E.Q.; formal analysis, M.L.; investigation, M.L., D.A.P.; resources, P.H.; data curation, F.P.; writing—original draft preparation, Co-First Authors, M.L., D.A.P., J.J.B.; writing—review and editing, J.J.B., G.A.G., P.H., F.P., and J.E.Q.; visualization, M.L. and D.A.P.; supervision, J.E.Q.; project administration, P.H. and J.E.Q.; funding acquisition, G.A.G., N.R.Z. All authors have read and agreed to the published version of the manuscript.

Funding: This research was funded by NIH CEBRA II DA017186. M.L. was supported by Wenner-Gren Foundation, Stockholm, Sweden.

Conflicts of Interest: GAG is principal owner of Quanteon LLC, J.E.Q., F.P., P.H., and J.J.B. serve as consultants to Quanteon LLC. The funders had no role in the design of the study; in the collection, analyses, or interpretation of data; in the writing of the manuscript; or in the decision to publish the results.

References

1. Siegel, G.J. *Basic Neurochemistry: Molecular, Cellular, and Medical Aspects*, 7th ed.; Elsevier Academic Press: Burlington, MA, USA, 2006.
2. Schultz, W. Multiple DA Functions at Different Time Courses. *Annu. Rev. Neurosci.* **2007**, *30*, 259–288. [CrossRef] [PubMed]
3. Perry, M.; Li, Q.; Kennedy, R.T. Review of Recent Advances in Analytical Techniques for the Determination of Neurotransmitters. *Anal. Chim. Acta* **2009**, *653*, 1–22. [CrossRef] [PubMed]
4. Grace, A.A. The Tonic/Phasic Model of DA System Regulation: Its Relevance for Understanding How Stimulant Abuse Can Alter Basal Ganglia Function. *Drug Alcohol Depend.* **1995**, *37*, 111–129. [CrossRef]
5. Cragg, S.J.; Rice, M.E. DAncing Past the DAT at a DA Synapse. *Trends Neurosci.* **2004**, *27*, 270–277. [CrossRef] [PubMed]
6. Watson, C.J.; Venton, B.J.; Kennedy, R.T. In Vivo Measurements of Neurotransmitters by Microdialysis Sampling. *Anal. Chem.* **2006**, *78*, 1391–1399. [CrossRef]
7. Jones, S.R.; Gainetdinov, R.R.; Caron, M.G. Application of Microdialysis and Voltammetry to Assess DA Functions in Genetically Altered Mice: Correlation with Locomotor Activity. *Psychopharmacology* **1999**, *147*, 30–32. [CrossRef]

8. Budygin, E.A.; Kilpatrick, M.R.; Gainetdinov, R.R.; Wightman, R.M. Correlation between Behavior and Extracellular DA Levels in Rat Striatum: Comparison of Microdialysis and Fast-Scan Cyclic Voltammetry. *Neurosci. Lett.* **2000**, *281*, 9–12. [CrossRef]
9. Parrot, S.; Bert, L.; Mouly-Badina, L.; Sauvinet, V.; Colussi-Mas, J.; Lambás-Señas, L.; Robert, F.; Bouilloux, J.P.; Suaud-Chagny, M.F.; Denoroy, L.; et al. Microdialysis Monitoring of Catecholamines and Excitatory Amino Acids in the Rat and Mouse Brain: Recent Developments Based on Capillary Electrophoresis with Laser-Induced Fluorescence Detection—A Mini-Review. *Cell. Mol. Neurobiol.* **2003**, *23*, 793–804. [CrossRef]
10. Clapp-Lilly, K.L.; Roberts, R.C.; Duffy, L.K.; Irons, K.P.; Hu, Y.; Drew, K.L. An Ultrastructural Analysis of Tissue Surrounding a Microdialysis Probe. *J. Neurosci. Methods* **1999**, *90*, 129–142. [CrossRef]
11. Borland, L.M.; Michael, A.C. *Electrochemical Methods for Neuroscience*; Michael, A.C., Borland, L.M., Eds.; CRC Press, Taylor and Francis Group: Boca Raton, FL, USA, 2007.
12. Mitala, C.M.; Wang, Y.; Borland, L.M.; Jung, M.; Shand, S.; Watkins, S.; Weber, S.G.; Michael, A.C.; Yang, H. In Vivo Fast-Scan Cyclic Voltammetry of DA near Microdialysis Probes. *J. Neurosci. Methods* **2008**, *174*, 177–185. [CrossRef]
13. Yang, H.; Peters, J.L.; Michael, A.C. Coupled Effects of Mass Transfer and Uptake Kinetics on In Vivo Microdialysis of DA. *J. Neurochem.* **1998**, *71*, 684–692. [CrossRef]
14. Clark, J.J.; Sandberg, S.G.; Wanat, M.J.; Gan, J.O.; Horne, E.A.; Hart, A.S.; Akers, C.A.; Parker, J.G.; Willuhn, I.; Martinez, V.; et al. Chronic Microsensors for Longitudinal, Subsecond DA Detection in Behaving Animals. *Nat. Methods* **2010**, *7*, 126–129. [CrossRef] [PubMed]
15. Robinson, D.L.; Heien, M.L.; Wightman, R.M. Frequency of DA Concentration Transients Increases in Dorsal and Ventral Striatum of Male Rats during Introduction of Conspecifics. *J. Neurosci.* **2002**, *22*, 10477–10486. [CrossRef] [PubMed]
16. Garris, P.A.; Ciolkowski, E.L.; Pastore, P.; Wightman, R.M. Efflux of DA from the Synaptic Cleft in the Nucleus Accumbens of the Rat Brain. *J. Neurosci.* **1994**, *14*, 6084–6093. [CrossRef] [PubMed]
17. Michael, A.C.; Borland, L.M.; Mitala, J.J.; Willoughby, B.M.; Motzko, C.M. Theory for the Impact of Basal Turnover on DA Clearance Kinetics in the Rat Striatum After Medial Forebrain Bundle Stimulation and Pressure Ejection. *J. Neurochem.* **2005**, *94*, 1202–1211. [CrossRef]
18. Zahniser, N.R.; Larson, G.A.; Gerhardt, G.A. In Vivo DA Clearance Rate in Rat Striatum: Regulation by Extracellular DA Concentration and DA Transporter Inhibitors. *J. Pharm. Exp. Ther.* **1999**, *289*, 266–277.
19. Gerhardt, G.A.; Burmeister, J.J. *Encyclopedia Analytical Chemistry: Instrumentation and Applications*; Meyers, R.A., Ed.; John Wiley and Sons: Chichester, UK, 2000; pp. 710–731.
20. Hascup, E.R.; af Bjerkén, S.; Hascup, K.N.; Pomerleau, F.; Huettl, P.; Strömberg, I.; Gerhardt, G.A. Histological Studies of the Effects of Chronic Implantation of Ceramic-Based Microelectrode Arrays and Microdialysis Probes in Rat Prefrontal Cortex. *Brain Res.* **2009**, *1291*, 12–20. [CrossRef]
21. Jaquins-Gerstl, A.; Michael, A.C. Comparison of the Brain Penetration Injury Associated with Microdialysis and Voltammetry. *J. Neurosci. Methods* **2009**, *183*, 127–135. [CrossRef]
22. Yang, C.; Cao, Q.; Puthongkham, P.; Lee, S.T.; Ganesana, M.; Lavrik, N.V.; Venton, B.J. 3D-Printed Carbon Electrodes for Neurotransmitter Detection. *Angew. Chem.* **2018**, *57*, 14255–14259. [CrossRef]
23. Cheng, Y.; Hu, K.; Wang, D.; Zubi, Y.; Lee, S.T.; Puthongkham, P.; Mirkin, M.V.; Venton, B.J. Cavity Carbon-Nanopipette Electrodes for Dopamine Detection. *Anal. Chem.* **2019**, *91*, 4618–4624. [CrossRef]
24. Schwerdt, H.N.; Shimazu, H.; Amemori, K.; Amemori, S.; Tierney, P.L.; Gibson, D.J.; Hong, S.; Yoshida, T.; Langer, R.; Cima, M.J.; et al. Chronic fast-scan dopamine voltammetry in primates. *Proc. Natl. Acad. Sci. USA* **2017**, *114*, 13260–13265. [CrossRef] [PubMed]
25. Liu, C.; Zhao, Y.; Cai, X.; Xie, Y.; Wang, T.; Cheng, D.; Li, L.; Li, R.; Deng, Y.; Ding, H.; et al. A wireless, implantable optoelectrochemical probe for optogenetic stimulation and dopamine detection. *bioRxiv* **2020**, *6*, 1–12. [CrossRef]
26. Schwarting, R.K.W.; Huston, J.P. The Unilateral 6-HydroxyDA Lesion Model in Behavioral Brain Research. Analysis of Functional Deficits, Recovery and Treatments. *Prog. Neurobiol.* **1996**, *50*, 275–331. [CrossRef]
27. Hascup, K.N.; Rutherford, E.C.; Quintero, J.E.; Day, B.K.; Nickell, J.R.; Pomerleau, F.; Huettl, P.; Burmeister, J.J.; Gerhardt, G.A. *Electrochemical Methods for Neuroscience*; Michael, A.C., Borland, L.M., Eds.; CRC Press: Boca Raton, FL, USA, 2007; pp. 407–450. [CrossRef]
28. Paxinos, G.; Watson, C. *The Rat Brain in Stereotaxic Coordinates*; Elsevier: Amsterdam, The Netherlands, 2007.

29. Burmeister, J.J.; Moxon, K.; Gerhardt, G.A. Ceramic-Based Multisite Microelectrodes for Electrochemical Recordings. *Anal. Chem.* **2000**, *72*, 187–192. [CrossRef]
30. Day, B.K.; Pomerleau, F.; Burmeister, J.J.; Huettl, P.; Gerhardt, G.A. Microelectrode Array Studies of Basal and Potassium-Evoked Release of L-glutamate in the Anesthetized Rat Brain. *J. Neurochem.* **2006**, *96*, 1626–1635. [CrossRef]
31. Gerhardt, G.A.; Oke, A.F.; Nagy, G.; Moghaddam, B.; Adams, R.N. Nafion-Coated Electrodes with High Selectivity for CNS Electrochemistry. *Brain Res.* **1984**, *290*, 390–395. [CrossRef]
32. Burmeister, J.J.; Gerhardt, G.A. Self-Referencing Ceramic-Based Multisite Microelectrodes for the Detection and Elimination of Interferences from the Measurement of L-glutamate and Other Analytes. *Anal. Chem.* **2001**, *73*, 1037–1042. [CrossRef]
33. Burmeister, J.J.; Pomerleau, F.; Huettl, P.; Gash, C.R.; Werner, C.E.; Bruno, J.P.; Gerhardt, G.A. Ceramic-Based Multisite Microelectrode Arrays for Simultaneous Measures of Choline and Acetylcholine in CNS. *Biosens. Bioelectr.* **2008**, *23*, 1382–1389. [CrossRef]
34. Hinzman, J.M.; Thomas, T.C.; Burmeister, J.J.; Quintero, J.E.; Huettl, P.; Pomerleau, F.; Gerhardt, G.A.; Lifshitz, J. Diffuse Brain Injury Elevates Tonic Glutamate Levels and Potassium-Evoked Glutamate Release in Discrete Brain Regions at Two Days Post-Injury: An Enzyme-Based Microelectrode Array Study. *J. Neurotrauma* **2010**, *27*, 889–899. [CrossRef]
35. Cass, W.A.; Zahniser, N.R.; Flach, K.A.; Gerhardt, G.A. Clearance of Exogenous DA in Rat Dorsal Striatum and Nucleus Accumbens: Role of Metabolism and Effects of Locally Applied Uptake Inhibitors. *J. Neurochem.* **1993**, *61*, 2269–2278. [CrossRef]
36. Friedemann, M.N.; Gerhardt, G.A. Regional Effects of Aging on DArgic Function in the Fischer-344 Rat. *Neurobiol. Aging* **1992**, *13*, 325–332. [CrossRef]
37. Hall, M.E.; Hoffer, B.J. Rapid and Sensitive Determination of Catecholamines in Small Tissues Samples by High Performance Liquid Chromatography Coupled with Dual-Electrode Coulometric electrochemical detection. *LCGC* **1989**, *7*, 258–265.
38. Herrera-Marschitz, M.; Goiny, M.; Utsumi, H.; Ungerstedt, U. Mesencephalic DA Innervation of the Frontoparietal (Sensorimotor) Cortex of the Rat: A Microdialysis Study. *Neurosci. Lett.* **1989**, *97*, 266–270. [CrossRef]
39. Schwarz, A.J.; Zocchi, A.; Reese, T.; Gozzi, A.; Garzotti, M.; Varnier, G.; Curcuruto, O.; Sartori, I.; Girlanda, E.; Biscaro, B.; et al. Concurrent Pharmacological MRI and In Situ Microdialysis of Cocaine Reveal a Complex Relationship between the Central hemodynamic response and Local DA Concentration. *Neuroimage* **2004**, *23*, 296–304. [CrossRef] [PubMed]
40. Pomerleau, F.; Day, B.K.; Huettl, P.; Burmeister, J.J.; Gerhardt, G.A. Real time in vivo measures of L-glutamate in the rat central nervous system using ceramic-based multisite microelectrode arrays. *Ann. N. Y. Acad. Sci.* **2003**, *1003*, 454–457. [CrossRef]
41. Van Horne, C.; Hoffer, B.J.; Strömberg, I.; Gerhardt, G.A. Clearance and Diffusion of Locally Applied DA in Normal and 6-HydroxyDA-Lesioned Rat Striatum. *J. Pharm. Exp. Ther.* **1992**, *263*, 1285–1292.
42. Hebert, M.A.; Gerhardt, G.A. Behavioral and neurochemical effects of intranigral administration of glial cell line-derived neurotrophic factor on aged Fischer 344 rats. *J. Pharmacol. Exp. Ther.* **1997**, *282*, 760–768.
43. Hascup, E.R.; Hascup, K.N.; Stephens, M.; Pomerleau, F.; Huettl, P.; Gratton, A.; Gerhardt, G.A. Rapid Microelectrode Measurements and the Origin and Regulation of Extracellular Glutamate in Rat Prefrontal Cortex. *J. Neurochem.* **2010**, *115*, 1608–1620. [CrossRef]
44. Rutherford, E.C.; Pomerleau, F.; Huettl, P.; Strömberg, I.; Gerhardt, G.A. Chronic Second-by-Second Measures of L-Glutamate in the Central Nervous System of Freely Moving Rats. *J. Neurochem.* **2007**, *102*, 712–722. [CrossRef]
45. Hascup, K.N.; Hascup, E.R.; Stephens, M.L.; Glaser, P.E.; Yoshitake, T.; Mathé, A.A.; Gerhardt, G.A.; Kehr, J. Resting Glutamate Levels and Rapid Glutamate Transients in the Prefrontal Cortex of the Flinders Sensitive Line Rat: A Genetic Rodent Model of Depression. *Neuropsychopharmacology* **2011**, *36*, 1769–1777. [CrossRef] [PubMed]
46. Burmeister, J.J.; Palmer, M.; Gerhardt, G.A. L-Lactate Measures in Brain Tissue with Ceramic-Based Multisite Microelectrodes. *Biosens. Bioelectron.* **2005**, *20*, 1772–1779. [CrossRef] [PubMed]
47. Tang, A.; Bungay, P.M.; Gonzales, R.A. Characterization of probe and tissue factors that influence interpretation of quantitative microdialysis experiments for dopamine. *J. Neurosci. Methods* **2003**, *126*, 1–11. [CrossRef]

48. Chen, N.H.; Lai, Y.J.; Pan, W.H. Effects of different perfusion medium on the extracellular basal concentration of dopamine in striatum and medial prefrontal cortex: A zero-net flux microdialysis study. *Neurosci. Lett.* **1997**, *225*, 197–200. [CrossRef]
49. Parsons, L.H.; Justice, J.B., Jr. Extracellular concentration and in vivo recovery of dopamine in the nucleus accumbens using microdialysis. *J. Neurochem.* **1992**, *58*, 212–218. [CrossRef] [PubMed]
50. Martin-Fardon, R.; Sandillon, F.; Thibault, J.; Privat, A.; Vignon, J. Long-term monitoring of extracellular dopamine concentration in the rat striatum by a repeated microdialysis procedure. *J. Neurosci. Methods* **1997**, *72*, 123–135. [CrossRef]
51. Taylor, I.M.; Patel, N.A.; Freedman, N.C.; Castagnola, E.; Cui, X.T. Direct in Vivo Electrochemical Detection of Resting Dopamine Using Poly(3,4-ethylenedioxythiophene)/Carbon Nanotube Functionalized Microelectrodes. *Anal. Chem.* **2019**, *91*, 12917–12927. [CrossRef]
52. Arnold, E.B.; Molinoff, P.B.; Rutledge, C.O. The release of endogenous norepinephrine and dopamine from cerebral cortex by amphetamine. *J. Pharmacol. Exp. Ther.* **1977**, *202*, 544–557.
53. Santos, R.M.; Lourenço, C.F.; Pomerleau, F.; Huettl, P.; Gerhardt, G.A.; Laranjinha, J.; Barbosa, R.M. Brain Nitric Oxide Inactivation Is Governed by the Vasculature. *Antioxid. Redox Signal.* **2010**, *14*, 1011–1021. [CrossRef]
54. Talauliker, P.M.; Price, D.A.; Burmeister, J.J.; Nagarid, S.; Pomerleau, F.; Huettl, P.; Hastings, J.T.; Gerhardt, G.A. Ceramic-Based Microelectrode Arrays: Recording Surface Characteristics and Topographical Analysis. *J. Neurosci. Methods* **2011**, *198*, 222–229. [CrossRef]
55. Burmeister, J.J.; Pomerleau, F.; Palmer, M.; Day, B.K.; Huettl, P.; Gerhardt, G.A. Improved ceramic-based multisite microelectrode for rapid measurements of L-glutamate in the CNS. *J. Neurosci. Methods* **2002**, *119*, 163–171. [CrossRef]
56. Burmeister, J.J.; Gerhardt, G.A. Ceramic-based multisite microelectrode arrays for in vivo electrochemical recordings of glutamate and other neurochemicals. *Trends Anal. Chem.* **2003**, *22*, 498–502. [CrossRef]
57. Bard, A.; Faulkner, L. *Electrochemical Methods: Fundamentals and Applications*, 2nd ed.; Wiley & Sons: Hoboken, NJ, USA, 2000.
58. Williams, J.M.; Galli, A. The dopamine transporter: A vigilant border control for psychostimulant action. *Handb. Exp. Pharmacol.* **2006**, *175*, 215–232. [CrossRef]

© 2020 by the authors. Licensee MDPI, Basel, Switzerland. This article is an open access article distributed under the terms and conditions of the Creative Commons Attribution (CC BY) license (http://creativecommons.org/licenses/by/4.0/).

Article

A Microfluidic Approach for Biosensing DNA within Forensics

Brigitte Bruijns [1,2,*,†], **Roald Tiggelaar** [1,3] **and Han Gardeniers** [1]

1. Mesoscale Chemical Systems, MESA+ Institute, University of Twente, Drienerlolaan 5, 7500 AE Enschede, The Netherlands; r.m.tiggelaar@utwente.nl (R.T.); j.g.e.gardeniers@utwente.nl (H.G.)
2. Life Sciences, Life Sciences, Engineering & Design, Saxion University of Applied Sciences, M. H. Tromplaan 28, 7513 AB Enschede, The Netherlands
3. NanoLab Cleanroom, MESA+ Institute, University of Twente, Drienerlolaan 5, 7500 AE Enschede, The Netherlands
* Correspondence: brigitte.bruijns@micronit.com
† Current address: Micronit Microtechnologies BV, Colosseum 15, 7521 PV Enschede, The Netherlands.

Received: 2 August 2020; Accepted: 30 September 2020; Published: 12 October 2020

Abstract: Reducing the risk of (cross-)contamination, improving the chain of custody, providing fast analysis times and options of direct analysis at crime scenes: these requirements within forensic DNA analysis can be met upon using microfluidic devices. To become generally applied in forensics, the most important requirements for microfluidic devices are: analysis time, method of DNA detection and biocompatibility of used materials. In this work an overview is provided about biosensing of DNA, by DNA profiling via standard short tandem repeat (STR) analysis or by next generation sequencing. The material of which a forensic microfluidic device is made is crucial: it should for example not inhibit DNA amplification and its thermal conductivity and optical transparency should be suitable for achieving fast analysis. The characteristics of three materials frequently used materials, i.e., glass, silicon and PDMS, are given, in addition to a promising alternative, viz. cyclic olefin copolymer (COC). New experimental findings are presented about the biocompatibility of COC and the use of COC chips for multiple displacement amplification and real-time monitoring of DNA amplification.

Keywords: biosensing; DNA analysis; forensics

1. Introduction

Sampling and securing traces at a crime scene is a crucial step in the police investigation process. Information obtained at this stage immediately gives direction to the investigation. Nowadays it takes weeks to get from collection of the evidence to a report provided by a forensic laboratory. By using microfluidic devices, also called chips, for carrying out (part of) the analysis directly at the crime scene, valuable information can be available at a much earlier stage in the investigation process. Such microfluidic devices are portable, need minimal analyte volumes and bring quick analysis times. Since sample handling in microfluidic devices is done in a sealed microfluidic environment, chips also improve the chain of custody and lower the risk of (cross-)contamination. As such, (forensic) interest in DNA analysis in microfluidic devices has arisen, with a preference for single use (disposable) chips.

Several systems have been developed that integrate the complete process from sampling to result in a single microfluidic device. The focus for these systems is on sequencing [1] or DNA amplification, often with the detection on-chip, but without the integration of sample preparation [2,3]. Commercial systems are available, although their deployment is still limited. For instance the RapidHIT produces a full short tandem repeat (STR) profile for 5–7 samples simultaneously in 90 min [4]. More recently the ANDE Rapid DNA system has been approved by the Federal Bureau of Investigation, but also with this machine it takes about 90 min from sample to result [5].

Having a fast analysis is of utmost importance at the scene of crime, for instance in the case of the identification of human remains or in sexual assault cases [6,7]. This is one of the motivations for the interest in microfluidic devices, which will be elaborated in the following section. Subsequently, two microfluidic DNA biosensing examples will be explored, namely DNA profiling by short tandem repeat (STR) analysis and by next generation sequencing (NGS). The biochemical compatibility of the materials used for microfluidic devices is an important requirement for forensic DNA analysis. Three commonly used materials (i.e., silicon, glass and PDMS) will be compared in terms of biocompatibility, thermal conductivity and optical transparency. The latter two characteristics play a role in fast amplification protocols and optical detection. An interesting alternative, cyclic olefin copolymer (COC), will be discussed on the basis of our own experimental findings, which demonstrates that this promising material can compete with reported materials for disposable forensic microfluidic devices. Finally, results regarding the use of COC microfluidic devices for multiple displacement amplification and on-chip real-time fast analysis of DNA amplification will be shown.

2. The Need for Speed

As published before by us [2], when performed at the forensic laboratory the conventional DNA analysis process might require days, as a consequence of which the outcome may have become less relevant for the initial phase of the criminal investigation as run by the police forces [8]. Such a time span may give a perpetrator the opportunity to remove relevant traces, to disappear or even to commit another crime [9]. The first hours of investigation are not without reason called the 'golden hours'. This motivates why there is a strong need for relevant information becoming available as quickly as possible [10]. Devices that supply police investigators at the crime scene with immediate information are especially valuable, since direct analysis assists fast and effective development of the case scenario.

Over the years devices for carrying out DNA analysis became more and more important as forensic evidence. In 1991, when the polymerase chain reaction (PCR) technique to amplify DNA was introduced within forensic research in the Netherlands, the minimal amount of DNA needed for a reliable result was 1 ng (corresponding to about 170 cells). Nowadays the amount necessary to obtain a complete and accurate STR profile is 100 pg. This means that even from a contact trace a reliable profile can be obtained, and only 1-10 cells are needed to obtain a complete DNA profile by using low copy number (LCN) PCR [11,12]. The DNA success rate highly depends on the DNA concentration; basically all traces with a concentration above 100 pg/µL result in DNA profiles that can be used for DNA database storage [13].

By the use of rapid identification technologies, like real-time analysis of DNA, investigators might identify a leading scenario early in the investigation process. This information can give direction to the further search for traces. Verification of a scenario can take place already at the crime scene and the scenario can be further reconstructed and adapted during the investigation process. De Gruijter et al. showed that receiving information concerning identification already at the crime scene helps to construct a more accurate scenario [14].

3. Biosensing of DNA

In this section various options regarding biosensing of DNA using the microfluidic approach. To generate DNA profiles STR analysis is the method of use will be described. Recently, with the introduction of sequencing techniques, also NGS methods are upcoming for forensic DNA profiling.

3.1. STR Analysis

STR analysis is the method within forensic science to generate DNA profiles. After amplification of the DNA, by PCR, the separation and detection of the amplified DNA must be carried out, which is usually accomplished by capillary electrophoresis (CE) [15].

Chips for PCR can be divided in two categories: well-based and continuous-flow chips [16]. In the first case, a well-based system (Figure 1A), the complete microdevice is sequentially cooled and heated. Drawback of a well-based system is the long time per cycle, caused by the relatively large total thermal mass of the system [16,17]. Continuous-flow chips can be further split up in three categories: fixed-loop, closed-loop and oscillatory chips. A fixed-loop microdevice (Figure 1B) contains zones with different temperatures through which the sample is moved. The numbers of meanders in the microdevice defines the number of thermal cycles. The time of each step is typically controlled by the length of the meander in a specific temperature zone in combination with the flow rate. In a closed-loop chip (Figure 1D) the circuit is fixed through which the sample is moved, while the number of thermal cycles can vary. Additionally, in an oscillatory microdevice the number of cycles can be varied (Figure 1C). The different zones of the microdevice are held at different temperatures and the sample is shunted back and forth between these different zones [16].

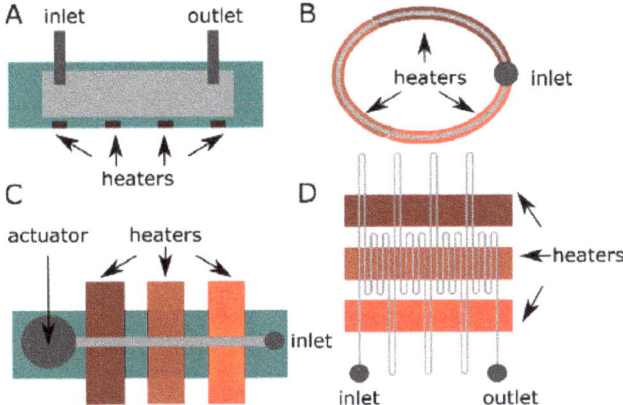

Figure 1. Examples of polymerase chain reaction (PCR) chip designs: (**A**) well-based (cross-section), (**B**) fixed-loop (top view), (**C**) oscillatory (top view) and (**D**) closed-loop systems (top view). The microfluidic amplification chamber (well-based) or channel (continous-flow) is given in grey. The temperature zones (continuous-flow) are given in the various shades of red.

In order to obtain a STR profile, in forensics there is significant interest in chips for CE to accomplish this [18]. By using such devices the required reagent and sample volumes are decreased, as well as the total analysis time [15]. Typical duration times of the separation are about 5–15 min [19–21].

3.2. Next Generation Sequencing

DNA sequencing is an essential technique within biotechnology, virology and medical diagnostics. In addition, for forensics these techniques have gained attention. In cases of a limited amount of DNA or highly degraded DNA the current NGS systems can help out.

The power of sequencing is that standard STR-profiles can be generated, but also DNA repeats can be sequenced to look for polymorphisms. To acquire information on ancestry, paternity or phenotype additional single-nucleotide polymorphisms (SNPs) can be analyzed with NGS techniques. However, the NGS workflow takes much longer than the conventional STR-profiling workflow (typically days versus hours).

There are several NGS machines on the market, such as the systems from Life Technologies, Illumina and PacBio. For forensic applications basically only the MiSeq/HiSeq (Illumina) and the Ion Torrent are used, with Nanopore as an upcoming sequencing system for forensics. A more elaborate review on NGS for forensics is given elsewhere [22].

3.2.1. MiSeq/HiSeq

Solexa, later on acquired by Illumina, made a sequencing system based on sequencing-by-synthesis in combination with reversible dye terminators and a planar solid glass support. The adapters are placed at the ends of the DNA sequence. [23–26]. Sequences complementary to the adapters 'A' and 'B' are present on the complete inside of these flow cell lanes [27]. Once the other end of the target DNA hybridizes to the complementary sequence present on the support, a bridge structure arises [24]. Each one of those bridge amplified clusters contains a unique DNA template, which is subsequently primed and sequenced [26]. There are different fluorescent labels present on each of the ddNTPs and a removable blocking group. The DNA sequence can be determined by completing the template one base at a time and recording of the fluorescent signal with a CCD camera. The HiSeq series, and later the MiSeq series, succeeded the Solexa Genome Analyzer [28].

3.2.2. Ion Torrent

The Ion Personal Genome Machine (PGM) was introduced by Ion Torrent at the end of 2010 [28,29]. This sequencing-by-synthesis method starts with emulsion PCR on beads. When a new nucleotide binds, upon pyrophosphate cleavage, a proton releases which is detected by monitoring the potential [28]. A well plate ensures that the release of these protons can be localized and retained. It is possible to sequence homopolymeric regions of the template DNA, since the signal is proportional to the amount of protons released. Data collection is done with a complementary metal-oxide semiconductor (CMOS) sensor array chip located at the bottom of the well plate. These chips can measure millions to billions of simultaneous sequencing reactions [30]. By measuring the potential, the system is faster, cheaper and accomplished with smaller instruments in comparison to systems that are based on fluorescence read-out [28].

3.2.3. Nanopore

A single DNA molecule can be read by the use of nanopores. This can be accomplished without the need for amplification or expensive fluorescent labels [31]. The detection principle of nanopore based systems depends on the ion current, which is generated when a charged molecule, such as a DNA molecule, passes through a nanoscale pore in a membrane resulting in a change of the ionic current. The four different bases produce four distinct current levels, which can be used for sequencing the DNA strand [31,32].

3.3. Conclusion

Flow cells, which are part of most NGS machines, have nanopatterned features in the microfluidic channels. These nanosized wells make it possible to obtain a high data density by patterned clustering of DNA fragments [33]. Microfluidic devices or flow cells can be used for STR analysis and for NGS, respectively. With NGS techniques more genetic information can be obtained besides the standard DNA profile, but the complete analysis process takes longer than standard STR analysis.

4. Biocompatibility of Materials for Forensic Microfluidic Devices

The material choice for (forensic) microfluidic chips is very critical. Each material has its own characteristics in the areas of, for instance, transparency, chemical resistance, thermal conductivity and certainly also biochemical compatibility, as shown in Table 1. Although more chip materials are available, such as poly(methyl methacrylate) (PMMA), polystyrene (PS), polyethylene terephthalate (PET) and polytetrafluoroethylene (PTFE). In this article only glass, silicon and the plastics PDMS and COC are considered, since these materials are mostly reported as chip materials for forensic microfluidic devices.

Table 1. Characteristics of different chip materials.

	Glass Borofloat	Silicon With Layer of Native Oxide	PDMS Sylgard 184	COC Topas 6013
Wettability Contact angle (H_2O)	Hydrophilic ≈ 35° [34]	Hydrophilic ≈ 57° [35]	Hydrophobic ≈ 105° [36]	Hydrophobic ≈ 88° [37–39]
Inhibitory (to amplification) Applied coating (to avoid inhibition)	Yes PEG 8000, PVP, EPDMA [47,48]	Yes SurfaSil, C_3H_9SiCl, $C_3H_7Cl_3Si$ [49,50]	Yes Parylene, BSA [40–44]	Yes BSA, PEG [45,46]
Transparency Light transmission (350–750 nm)	Transparant ≈ 91% [51]	Opaque [17] 0%	Transparant ≈ 90% [52]	Transparant ≈ 91% [39,53]
Thermal conductivity $W/m \cdot K$	Medium 1.2 (90 °C) [51,54]	High 156 (300 K) [55]	Low 0.27 (unknown K) [54,56]	Low 0.12–0.15 (20 °C) [53]
Chemical resistance Affected by	High HF, NaOH [51]	High KOH [57]	Low Chloroform, benzene, ethanol, acetone [58,59]	High Non-polar organic solvents [39,53]

63

By using microfluidic devices for the amplification reaction, instead of polypropylene tubes or well plates in combination with a conventional thermocycler, an enormous gain in time is achieved. Forensic chips are for single-use only, i.e., disposable, and therefore these chips can be fabricated of a cheap base-material, provided that it meets all requirements. However, material selection is not always simple and easy, because the material itself can disadvantageously influence (some of) the steps in the process of DNA analysis, such as PCR and detection.

Biochemical compatibility of materials for forensic chips is of great importance, since an interaction of the used chemicals with the walls of a microfluidic channel can result in slowing down of the reaction or even total inhibition. The high surface-to-volume ratio of microfluidic devices is a disadvantage in this case. However, such unwanted interaction between the chip material and used chemicals can be avoided by either passivation of the walls of the reaction chamber prior to use, called passive coating, or by so-called dynamic coating by adding components to the used chemicals, e.g., in the amplification mixture [60].

Besides biochemical compatibility, also the wettability of the chip material is of great importance. Microfluidic devices for droplet PCR are usually made of PDMS [61–66], since the interior of a PDMS channels is hydrophobic and aqueous droplets-in-oil are easier generated in channels of which the surface has this wetting state (compared to a hydrophilic state of the surface of a microfluidic channel). In case of water-in-oil droplets, the template DNA that is present in the aqueous phase cannot adsorb to the interior of the fluidic channel because of the hydrophobic property of the oil [67]. On the other hand, the PCR mixture, as well as other reaction mixtures used for DNA analysis (e.g., lysing agents), are aqueous solutions. By using hydrophilic channel walls it is easier to introduce these kind of solutions into the chip without bubble formation [17].

Since the fabrication of microfluidic channels in glass and silicon often requires cleanroom facilities, due to which the production costs for such chips are relatively high, chips made of these materials are not very suitable as disposables (i.e., for single use only) [17]. Plastic devices can be made using e.g., hot embossing, micromilling, casting or injection molding, for which a cleanroom environment is not necessary, which makes such devices cheaper. Due to their relatively low price plastic chips are (most) appropriate for single-use (i.e., disposable), which in particular circumvents cross-contamination issues [68,69].

Whereas plastic chips are attractive because of their bio(chemical)compatibility, another advantage of this material is its low thermal conductivity [70]. Materials with a low thermal conductivity are desirable for continuous-flow devices for PCR (which requires different temperature zones in the chip), for which reason plastic as well as glass are attractive [68]. As such, the majority of chips for analysis of biological fluids is fabricated from PDMS or PMMA [64,71–76], although glass and silicon are also occasionally utilized [77–79].

The next subsections contain detailed information about various aspects, such as transparency, thermal conductivity and bio(chemical)compatibility of glass, silicon, COC and PMDS as chip material for forensic microfluidic devices.

4.1. Glass

Due to its transparency in the visible range, which gives opportunities for optical detection, glass is widely used a material for (forensic) DNA analysis on-chip. PCR and CE can be easily integrated on-chip due to the beneficial electro-osmotic-flow/electrical characteristics of glass [17,80,81].

For glass surfaces it is reported that silanization of the surface is required to minimise surface adsorption of Taq polymerase. Giordano et al. investigated the use of dynamic coating (polyethylene glycol (PEG) 8000, polyvinylpyrrolidone (PVP) and hydroxyethylcellulose (HEC)) versus passive (or static) coating (epoxy (poly)dimethylacrylamide (EPDMA)) of the reaction chamber of a PCR glass chip. Covalent silanization results in 50–120% of product relative to the amount obtained with conventional PCR in polypropylene tubes. The use of HEC ends in total inhibition of the reaction, whereas the use of PEG 8000, PVP and EPDMA results in 13%, 78% and 72% of relative product amount, respectively [48]. Other coatings that can be used to passivate the glass surface are SigmaCote [82] or dichlorodimethylsilane [83].

4.2. Silicon

Similar to glass, also silicon is frequently used as chip material, because of its high thermal conductivity. Due to the latter silicon is attractive for the fast heating and cooling as required for (well-based) PCR cycling [17]. Since silicon is a semiconductor, disadvantage of this material is that it cannot resist the high potentials needed for CE. Moreover, silicon is not transparent for the wavelengths of UV-Vis light, for which reason UV-Vis cannot be used for detection [70,84]. Based on a paper of Cho et al. [70], it is concluded that the majority of well-based PCR chips is made from silicon.

In fact, bare silicon shows inhibitory effects on the PCR reaction, which cannot be circumvented with the deposition of a silicon nitride layer. A silanizing agent followed by a polymer coating (e.g., polyglycine or polymaleimide) prevents inhibition, whereby SurfaSil shows better results than SigmaCote. An oxidised silicon surface can also prevent inhibition of the PCR reaction [49]. Felbel et al. investigated chlorotrimethylsilane, dichlorodimethylsilane, hexamethyldisilazane and trichloropropylsilane as silanization agents, which all prevent inhibition of the PCR reaction [50].

4.3. PDMS

PDMS is also reported as chip material for the DNA amplification reaction [4,85,86]. PDMS is stable over a temperature range of -50–$200\ °C$ [87]. Moreover, PDMS is biochemically compatible, easy to shape (by molding) and optically transparent [59,88]. PDMS is very suitable for cell culturing, since the permeability for gases, like O_2 and CO_2, is higher than for any other polymer [39]. However, PDMS suffers from severe bubble formation during the thermal cycling protocol, due to evaporation of the PCR mixture (through the PDMS) at temperatures above $90\ °C$ [40,89]. PDMS is not very compatible with common solvents, such as chloroform, benzene, acetone and ethanol, since PDMS swells in these chemicals [58,59]. Furthermore, it is reported that surface treatments on PDMS are often unstable over time [39].

Although PDMS is used as coating for glass PCR chips [90,91], an important drawback of the use of uncoated PDMS is that, because of its permeability, it might result in inhibition of the amplification reaction due to absorption or adsorption of components in the amplification mixture (in particular the enzyme). In fact, this drawback can be bypassed by coating the PDMS with bovine serum albumin (BSA) [42–44]. To coat PDMS chips, Shin et al. and Prakash et al. used a parylene (dichlorodiparaxylylene) surface treatment [40,41]. PVP is also applied by Kim et al. to coat a PDMS/glass PCR chip [92].

4.4. COC

COC, nowadays frequently applied in disposables for medical diagnostics as well as for food packaging, is a promising and interesting material for chips [39,53,93]. COC is a copolymer based on linear and cyclic olefins. Since some manufacturers make COC from more than one type of monomer the 'copolymer' part is included in the name [39].

By means of a variety of techniques microfluidic networks can be realized in COC, such as injection molding, micromilling and hot embossing. The COC grade determines the exact heat deflection temperature, which lays in the range 70–170 $°C$ [53,94]. Post-processing, two pieces of COC can be joint using thermal bonding, solvent bonding or gluing [39,93]. COC has a high rigidity and is optically transparent in the UV-range. Moreover, its water absorptivity is low (<0.01%) and it is electrically non-conductive. COC withstands a variety of chemicals (e.g., HCl, H_2SO_4, HNO_3, NaOH, EtOH and $(CH_3)_2CO$), and the material is only affected by some non-polar solvents (e.g., C_6H_{14} and C_7H_8) [39].

Plastic chips, either PMMA or COC, can be coated with a PEG solution [45]. A chip made from cyclic olefin polymer (COP) with a glass transition temperature of 138 $°C$ is coated with BSA prior to the PCR by Koh et al. [46]. Although BSA is widely used as surface treatment to prevent PCR inhibition, it might influence fluorescence detection, due to the possible interaction between BSA and the fluorescent dye or probe [17]. Panaro et al. tested the influence of dynamic coating on inhibition by adding 0.75% w/v PEG 8000 to the PCR mixture. For a large amount of plastics the addition of PEG 8000 improves the concentration of PCR product [47].

4.5. Conclusion

Silicon is a suitable material for well-based PCR systems, due to its good thermal conductivity. Glass or polymers, on the other hand, have a low thermal conductivity, making these chip materials suitable for continuous-flow PCR systems. Polymers, like COC, are also attractive for microfluidic devices for forensic DNA analysis due to their bio(chemical)compatibility, transparency and the the relatively wide variety of available fabrication techniques, which makes it possible to fabricate polymer-based chips at relatively low costs, making them suitable as disposables.

5. COC Chips for Multiple Displacement Amplification (MDA): Biocompatibility and Real-Time Analysis

As discussed in the previous section, COC is an attractive material for forensic micofluidic devices, since it has excellent bio(chemical)compatibility and chips of this material can be made at relatively low costs. This motivates our experimental investigation of COC for multiple displacement amplification (MDA) in microfluidic devices. It is noted that parts of the presented work are reported elsewhere as well by us [95–97].

Uniform amplification of small amounts of DNA is important for single-cell genomics, sequencing and forensic science. Due to amplification bias, the isothermal amplification method MDA is known for uneven representation of the template. Besides a reduction in the amount of reagents and sample, lowering the volume of the amplification reaction contributes to a more uniform amplification.

The compatibility of COC with the MDA amplification reaction is verified by performing an on-chip reaction. Two different designs of COC chips are investigated and chips with and without BSA treatment are used to investigate the need for a coating. In SolidWorks both designs are made. Design I contains a serpentine-shaped channel (width as well as depth 1.0 mm) with a volume of ca. 55 µL (Figure 2A), whereas design II comprises a straight channel (depth 0.5 mm, width 1.0 mm) with an internal volume of ca. 5 µL (Figure 2B). In the vicinity of the reaction channel a channel is present for insertion of a thermocouple that is used for temperature measurement and control. Both microfluidic layouts are milled in 2 mm thick (design I) or 1 mm thick (design II) COC plates (DENZ Bio-Medical grade 6013) using a Sherline 5410 Deluxe Mill with a 1.0 mm diameter mill, followed by drilling of fluidic accesses with a 1.5 mm diameter drill into which standard pipette tips can be mounted.

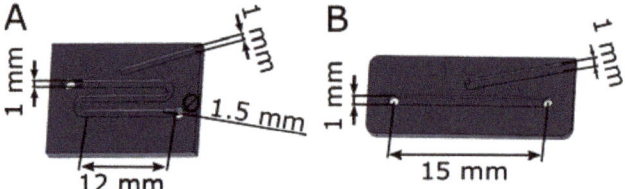

Figure 2. The schematic representation of the chips with the channel dimensions of (**A**) design I and (**B**) design II.

Post to milling and drilling, the COC chips are extensively rinsed with MilliQ water and ultrasonication (15 min) in ethanol, followed by nitrogen gas blow-drying. In order to recover the optical transparency, the surface roughness of the milled structures is lowered by following the procedure of Ogilvie et al. [98]: the chips are exposed (1 min) to cyclohexane vapor (60 °C).

Subsequently, the channel of each COC-chip is sealed by applying a piece of PCR foil (Microseal 'B' Adhesive Seals, Bio-Rad), which is a robust seal that is appropriate for biological applications [99].

Prior to performing the MDA reaction, some chips of design I are coated with BSA (1% (10 µg/µL)). The MDA reactions (GenomiPhi V2 kit from GE Healthcare) are carried out in the COC chips that are incubated in a water bath of 32 °C for 2h. The manufacturer of the MDA kit states that this kit is optimized for whole genome amplification from at least 10 ng [100]. Next to that an input concentration of 10 ng/µL is used, since off-chip results showed an enormous increase in the DNA concentration within 2 h [97]. It is noted that in case of design I 10 µL of the MDA mix (reaction buffer, sample buffer, enzyme mix, DNA and MilliQ in a ratio of 9:5:1:1:4, respectively) is injected in each chip, whereas the injected volume is 5 µL in case of design II. Post to loading of the indicated volume of MDA mixture, the fluidic accesses of the chips are closed with PCR foil. In addition, 10 µL MDA mixture is pipetted into a 500 µL Eppendorf polypropylene (PP) vial as a control. The concentration of double-stranded DNA (dsDNA) is determined with the Qubit™ dsDNA HS Assay Kit and the fluorescence is measured using the Qubit™ 3.0 fluorometer (both from Thermo Fisher Scientific). All samples are measured in triplicate.

Table 2 summarizes the outcome of the DNA quantification of the experiments with chip design I. As can be seen, the concentration of DNA in uncoated chips is well above 200 ng/µL (after a 2 h run), which is in agreement with the control vials, runs in a thermocycler (see [95,97] for more details) and literature [100–102]. Remarkable is that use of a BSA coating in the COC microfluidic channels clearly negatively affects the MDA reaction, in fact it results in inhibition, whereas such coating is known as method to prevent non-specific binding to COC surfaces [46,103–105].

Table 2. The concentration of double-stranded DNA (dsDNA) obtained after 2 h of amplification in the Eppendorf vials and chips of design I with and without BSA coating.

	Run 1 (ng/µL)	Run 2 (ng/µL)	Run 3 (ng/µL)	Run 4 (ng/µL)	Average [a] (ng/µL)
Eppendorf vial	204 ± 7	218 ± 8	265 ± 13	200 ± 16	222 ± 30
Chip without BSA coating	118 ± 6	209 ± 6	295 ± 13	233 ± 20	214 ± 73
Chip with BSA coating	79 ± 3	[b]	86 ± 2	62 ± 3	76 ± 12

[a] The standard deviation was taken from the average values of the four runs. [b] The seal of the chip was not secured properly and came off, hence no quantification measurement could be carried out.

The adsorption of polymerase on the channel walls can be prevented by the addition of BSA to the amplification mix or by coating the channel walls with BSA [106–108]. Erill et al. suggest to use a concentration of 2.5 µg/µL (0.25%) [107]. Taylor et al. state that high BSA concentrations (>0.15%) result in a lower yield compared to absence of BSA in the amplification mixture and that increasing the BSA concentration will finally result in total inhibition of the PCR reaction. The optimum BSA concentration for their silicon chip is found to be to be 0.05% resulting in a 2-fold increase in yield [108]. Kodzius et al. conclude that 2 µg/µL (0.2%) of BSA does not have any negative influence on the PCR reaction [106]. Kreader et al. recommend to use 0.2–0.4 µg/µL (0.02–0.04%) to relieve the inhibition of humic acids in PCR [109]. Qin et al. studied the optimal pH and BSA concentration to coat their Norland Optical Adhesive (NOA) chip. They report that the absorption capacity of the BSA to the NOA surface does not increase any further after 1 h. An acidic environment (pH 4) of the BSA solution shows a higher PCR efficiency than neutral pH. A BSA concentration of 5 µg/µL (0.5%) gives the best results. Using a lower concentration might not be sufficient, while a higher concentration can lead to PCR inhibition [110]. According to Zhang et al. [17], another reason for obtaining a lower DNA concentration upon using a BSA coating could be that salts present in the amplification mixtures (e.g., KCl, $MgCl_2$, $MgSO_4$ or NaOH) negatively affect the coating.

By using X-ray photo electron spectroscopy and contact angle measurements Sweryda-Krawiec et al. determined the mode of surface passivation by BSA. On a hydrophilic substrate BSA surface passivation shows two modes: it covers the surface and it can interact with an initially deposited layer for further adsorption. Surface passivation of a hydrophobic surface is reached in one step. The measured contact angle is the same after completion of the adsorption reaction [111]. Additionally, the BSA surface coverage is at maximum 53% for a complete monolayer on hydrophobic surfaces, while on hydrophilic surfaces an almost full monolayer can be present [112]. COC is hydrophobic, probably resulting in an incomplete monolayer of BSA.

In this research 10 µg/µL (1%) is used, which is higher than recommended concentrations [106,108,109], which probably causes (partial) inhibition of the MDA reaction. Additionally, the chip is coated with BSA (passive coating/static passivation) instead of adding the BSA solution to the amplification mixture (active coating/dynamic passivation), as a consequence of which the BSA can compete with the polymerase for adsorption, leading to a lower amplification yield.

The amplification yield of chip design I is compared with that of design II (Table 3), without the use of any coating. The on-chip amplification with design I shows similar results as shown in Table 2 and the control run in the Eppendorf vial (215 ng/µL). However, the chips with the smaller internal volume, design II, shows a lower amplification yield in comparison with design I.

Table 3. The concentration of dsDNA obtained after 2 h of amplification with chips of both designs, without BSA coating.

	Run 1 (ng/µL)	Run 2 (ng/µL)	Run 3 (ng/µL)	Run 4 (ng/µL)	Average [a] (ng/µL)
Chip design I	201 ± 4	262 ± 5	231 ± 35	215 ± 6	228 ± 26
Chip design II	207 ± 4	181 ± 0	185 ± 5	141 ± 4	179 ± 28

[a] The standard deviation was taken from the average values of the four runs.

Although lowering the amplification volume gives better amplification in terms of coverage, bias and specificity [113–117], the yield is not necessarily higher as shown in these experiments. The interaction of the polymerase with the channel walls increases upon a higher surface-to-volume ratio [106]. Chip design I and chip design II have a 6:1 and 4:1 surface-to-volume ratio, respectively. This 1.5 times larger surface-to-volume ratio of design II is assumed to be the reason for the lower amplification yield of the chips with design II.

In conclusion, for COC chips with a sufficiently low surface-to-volume ratio (4:1 or lower) there is no need for (BSA) coating. In case BSA coating is required to make the chip material bio(chemical)compatible the concentration of BSA has to be dosed to an optimum value.

In fact, the presented COC chips can be used for fast analysis. More specific, design I can be used for real-time monitoring of DNA amplification with the MDA reaction down to an input concentration of 0.01 ng/µL. In COC chips of design I, the MDA reaction is performed with different DNA concentrations. All required measurement and control functionality, i.e., temperature sensors and photosensors based on amorphous silicon, thin film metallic heaters as well as an interference filter, for monitoring the reaction in real-time (based on measuring the fluorescence intensity) are realized onto a so-called system-on-glass. The COC chip is positioned on top of this system-on-glass, making the setup a compact transportable system, with the potential to be used at a crime scene. The experimental data indicate that upon using on-chip real-time detection the presence of DNA in a sample can be observed within approximately 10 minutes [96], i.e., one only has to wait until the signal exceeds the threshold [118].

6. Conclusions

Microfluidic devices, by having a closed microfluidic network, improve the chain of custody, reduce the risk of (cross-)contamination and offer fast analysis times, making them very suitable to be used directly at the crime scene. By choosing the correct chip material such microfluidic devices offer the required optical transparency, biocompatibility and thermal conductivity. COC is a bio(chemical)compatible material for which no coating of the microfluidic channel walls is required (provided that fluidic network has a sufficiently low surface-to-volume ratio) and in which microfluidic channels can be made at relatively low costs, making them very suitable for single use only for forensic applications like DNA analysis.

Author Contributions: Conceptualization, B.B.; methodology, B.B.; investigation, B.B.; writing—original draft preparation, B.B. and R.T.; writing—review and editing, B.B., R.T. and H.G.; supervision, R.T. and H.G.; project administration, H.G.; funding acquisition, B.B. All authors have read and agreed to the published version of the manuscript.

Funding: This research received no external funding.

Conflicts of Interest: The authors declare no conflict of interest.

Abbreviations

The following abbreviations are used in this manuscript:

BSA	bovine serum albumin
CCD	charge coupled device
CE	capillary electrophoresis
CMOS	complementary metal-oxide semiconductor
COC	cyclic olefin copolymer
COP	cyclic olefin polymer
dsDNA	double-stranded DNA
DNA	deoxyribonucleic acid
EPDMA	epoxy (poly)dimethylacrylamide
HEC	hydroxyethylcellulose
LCN	low copy number
MDA	multiple displacement amplification
NFI	Netherlands Forensic Institute
NGS	next generation sequencing
NOA	Norland Optical Adhesive
PCR	polymerase chain reaction
PDMS	polydimethylsiloxane
PEG	polyethylene glycol
PET	polyethylene terephthalate
PMMA	poly(methyl methacrylate)
PP	polypropylene
PTFE	polytetrafluoroethylene
PS	polystyrene
PVP	polyvinylpyrrolidone
SNP	single-nucleotide polymorphism
STR	short tandem repeat

References

1. Tytgat, O.; Gansemans, Y.; Weymaere, J.; Rubben, K.; Deforce, D.; Van Nieuwerburgh, F. Nanopore Sequencing of a Forensic STR Multiplex Reveals Loci Suitable for Single-Contributor STR Profiling. *Genes* **2020**, *11*, 381. [CrossRef]
2. Bruijns, B.; Van Asten, A.; Tiggelaar, R.; Gardeniers, H. Microfluidic devices for forensic DNA analysis: A review. *Biosensors* **2016**, *6*, 41. [CrossRef]
3. Cornelis, S.; Tytgat, O.; Fauvart, M.; Gansemans, Y.; Vander Plaetsen, A.S.; Wiederkehr, R.S.; Deforce, D.; Van Nieuwerburgh, F.; Stakenborg, T. Silicon μPCR Chip for Forensic STR Profiling with Hybeacon Probe Melting Curves. *Sci. Rep.* **2019**, *9*, 1–9. [CrossRef] [PubMed]
4. Yang, J.; Hurth, C.; Nordquist, A.; Smith, S.; Zenhausern, F. Integrated Microfluidic System for Rapid DNA Fingerprint Analysis: A Miniaturized Integrated DNA Analysis System (MiDAS)—Swab Sample-In to DNA Profile-Out. In *Microfluidic Electrophoresis*; Springer: Berlin, Germany, 2019; pp. 207–224.
5. Ragazzo, M.; Melchiorri, S.; Manzo, L.; Errichiello, V.; Puleri, G.; Nicastro, F.; Giardina, E. Comparative Analysis of ANDE 6C Rapid DNA Analysis System and Traditional Methods. *Genes* **2020**, *11*, 582. [CrossRef] [PubMed]
6. Turingan, R.; Brown, J.; Kaplun, L.; Smith, J.; Watson, J.; Boyd, D.; Steadman, D.; Selden, R. Identification of human remains using Rapid DNA analysis. *Int. J. Legal Med.* **2020**, *134*, 863–872. [CrossRef]

7. Wolf, K.; Pakulla-Dickel, S.; del Río, A.G.; Prochnow, A.; Elliott, K.; Scherer, M. How the Investigator Casework GO! Kit provides sensitive, fast and robust direct amplification of low copy number samples. *Forensic Sci. Int. Genet. Suppl. Ser.* **2019**, *7*, 626–628. [CrossRef]
8. Mapes, A.; Kloosterman, A.; Poot, C. DNA in the criminal justice system: The DNA success story in perspective. *J. Forensic Sci.* **2015**, *60*, 851–856. [CrossRef]
9. van Asten, A. On the added value of forensic science and grand innovation challenges for the forensic community. *Sci. Justice* **2014**, *54*, 170–179. [CrossRef] [PubMed]
10. Kloosterman, A.; Mapes, A.; Geradts, Z.; van Eijk, E.; Koper, C.; van den Berg, J.; Verheij, S.; van der Steen, M.; van Asten, A. The interface between forensic science and technology: how technology could cause a paradigm shift in the role of forensic institutes in the criminal justice system. *Philos. Trans. B* **2015**, *370*, 20140264. [CrossRef]
11. Meulenbroek, A.; Kloosterman, A. DNA-onderzoek van minimale biologische sporen; gevoelige problematiek. *Analyse* **2009**, *64*, 108–120.
12. Kloosterman, A.; Kersbergen, P. Efficacy and limits of genotyping low copy number DNA samples by multiplex PCR of STR loci. *Int. Cong. Ser.* **2003**, *1239*, 795–798. [CrossRef]
13. Mapes, A.; Kloosterman, A.; Marion, V.; Poot, C. Knowledge on DNA success rates to optimize the DNA analysis process: from crime scene to laboratory. *J. Forensic Sci.* **2016**, *61*, 1055–1061. [CrossRef]
14. de Gruijter, M.; de Poot, C.; Elffers, H. Reconstructing with trace information: Does rapid identification information lead to better crime reconstructions? *J. Investig. Psychol. Offender Profil.* **2017**, *14*, 88–103. [CrossRef]
15. Han, J.; Sun, J.; Wang, L.; Liu, P.; Zhuang, B.; Zhao, L.; Liu, Y.; Li, C. The Optimization of Electrophoresis on a Glass Microfluidic Chip and its Application in Forensic Science. *J. Forensic Sci.* **2017**, *62*, 1603–1612. [CrossRef]
16. Zhang, Y.; Ozdemir, P. Microfluidic DNA amplification: A review. *Anal. Chim. Acta* **2009**, *638*, 115–125. [CrossRef]
17. Zhang, C.; Xing, D. Miniaturized PCR chips for nucleic acid amplification and analysis: latest advances and future trends. *Nucleic Acids Res.* **2007**, *35*, 4223–4237. [CrossRef]
18. Pascali, J.; Bortolotti, F.; Tagliaro, F. Recent advances in the application of CE to forensic sciences, an update over years 2009–2011. *Electrophoresis* **2012**, *33*, 117–126. [CrossRef]
19. Le Roux, D.; Root, B.; Reedy, C.; Hickey, J.; Scott, O.; Bienvenue, J.; Landers, J.; Chassagne, L.; de Mazancourt, P. DNA analysis using an integrated microchip for multiplex PCR amplification and electrophoresis for reference samples. *Anal. Chem.* **2014**, *86*, 8192–8199. [CrossRef]
20. Lagally, E.; Emrich, C.; Mathies, R. Fully integrated PCR-capillary electrophoresis microsystem for DNA analysis. *Lab Chip* **2001**, *1*, 102–107. [CrossRef]
21. Liu, P.; Li, X.; Greenspoon, S.; Scherer, J.; Mathies, R. Integrated DNA purification, PCR, sample cleanup, and capillary electrophoresis microchip for forensic human identification. *Lab Chip* **2011**, *11*, 1041–1048. [CrossRef]
22. Bruijns, B.; Tiggelaar, R.; Gardeniers, H. Massively parallel sequencing techniques for forensics: A review. *Electrophoresis* **2018**, *39*, 2642–2654. [CrossRef]
23. Van Dijk, E.L.; Auger, H.; Jaszczyszyn, Y.; Thermes, C. Ten years of next-generation sequencing technology. *Trends Genet.* **2014**, *30*, 418–426. [CrossRef]
24. Ansorge, W.J. Next-generation DNA sequencing techniques. *New Biotechnol.* **2009**, *25*, 195–203. [CrossRef]
25. Glenn, T.C. Field guide to next-generation DNA sequencers. *Mol. Ecol. Resour.* **2011**, *11*, 759–769. [CrossRef]
26. Stranneheim, H.; Lundeberg, J. Stepping stones in DNA sequencing. *Biotechnol. J.* **2012**, *7*, 1063–1073. [CrossRef]
27. Buermans, H.; Den Dunnen, J. Next generation sequencing technology: Advances and applications. *Biochim. Biophys. Acta BBA Mol. Basis Dis.* **2014**, *1842*, 1932–1941. [CrossRef]
28. Liu, L.; Li, Y.; Li, S.; Hu, N.; He, Y.; Pong, R.; Lin, D.; Lu, L.; Law, M. Comparison of next-generation sequencing systems. *BioMed Res. Int.* **2012**, *2012*. [CrossRef]
29. Heather, J.M.; Chain, B. The sequence of sequencers: The history of sequencing DNA. *Genomics* **2016**, *107*, 1–8. [CrossRef]
30. Merriman, B.; R&D Team, I.T.; Rothberg, J.M. Progress in ion torrent semiconductor chip based sequencing. *Electrophoresis* **2012**, *33*, 3397–3417. [CrossRef]

31. Schneider, G.F.; Dekker, C. DNA sequencing with nanopores. *Nat. Biotechnol.* **2012**, *30*, 326. [CrossRef]
32. Venkatesan, B.M.; Bashir, R. Nanopore sensors for nucleic acid analysis. *Nat. Nanotechnol.* **2011**, *6*, 615. [CrossRef]
33. Micronit Microtechnologies BV. Next-Generation Sequencing. 2020. Available online: https://www.micronit.com/microfluidics/next-generation-sequencing.html (accessed on 24 June 2020).
34. Vicente, C.; André, P.; Ferreira, R. Simple measurement of surface free energy using a web cam. *Revista Brasileira de Ensino de Física* **2012**, *34*, 1–5. [CrossRef]
35. Extrand, C.; Kumagai, Y. An experimental study of contact angle hysteresis. *J. Colloid Interface Sci.* **1997**, *191*, 378–383. [CrossRef]
36. Haubert, K.; Drier, T.; Beebe, D. PDMS bonding by means of a portable, low-cost corona system. *Lab Chip* **2006**, *6*, 1548–1549. [CrossRef]
37. Stachowiak, T.; Mair, D.; Holden, T.; Lee, L.; Svec, F.; Fréchet, J. Hydrophilic surface modification of cyclic olefin copolymer microfluidic chips using sequential photografting. *J. Sep. Sci.* **2007**, *30*, 1088–1093. [CrossRef]
38. Roy, S.; Yue, C.; Lam, Y.; Wang, Z.; Hu, H. Surface analysis, hydrophilic enhancement, ageing behavior and flow in plasma modified cyclic olefin copolymer (COC)-based microfluidic devices. *Sens. Actuators B Chem.* **2010**, *150*, 537–549. [CrossRef]
39. Nunes, P.; Ohlsson, P.; Ordeig, O.; Kutter, J. Cyclic olefin polymers: emerging materials for lab-on-a-chip applications. *Microfluid. Nanofluidics* **2010**, *9*, 145–161. [CrossRef]
40. Shin, Y.; Cho, K.; Lim, S.; Chung, S.; Park, S.J.; Chung, C.; Han, D.C.; Chang, J. PDMS-based micro PCR chip with parylene coating. *J. Micromech. Microeng.* **2003**, *13*, 768. [CrossRef]
41. Prakash, A.; Adamia, S.; Sieben, V.; Pilarski, P.; Pilarski, L.; Backhouse, C. Small volume PCR in PDMS biochips with integrated fluid control and vapour barrier. *Sens. Actuators B Chem.* **2006**, *113*, 398–409. [CrossRef]
42. Crabtree, H.; Lauzon, J.; Morrissey, Y.; Taylor, B.; Liang, T.; Johnstone, R.; Stickel, A.; Manage, D.; Atrazhev, A.; Backhouse, C.; et al. Inhibition of on-chip PCR using PDMS–glass hybrid microfluidic chips. *Microfluid. Nanofluidics* **2012**, *13*, 383–398. [CrossRef]
43. Niu, Z.; Chen, W.; Shao, S.; Jia, X.; Zhang, W. DNA amplification on a PDMS–glass hybrid microchip. *J. Micromech. Microeng.* **2006**, *16*, 425. [CrossRef]
44. Fernández-Carballo, B.; McGuiness, I.; McBeth, C.; Kalashnikov, M.; Borrós, S.; Sharon, A.; Sauer-Budge, A. Low-cost, real-time, continuous flow PCR system for pathogen detection. *Biomed. Microdevices* **2016**, *18*, 34.
45. Münchow, G.; Dadic, D.; Doffing, F.; Hardt, S.; Drese, K.S. Automated chip-based device for simple and fast nucleic acid amplification. *Expert Rev. Mol. Diagn.* **2005**, *5*, 613–620. [CrossRef]
46. Koh, C.; Tan, W.; Zhao, M.Q.; Ricco, A.; Fan, Z. Integrating polymerase chain reaction, valving, and electrophoresis in a plastic device for bacterial detection. *Anal. Chem.* **2003**, *75*, 4591–4598. [CrossRef]
47. Panaro, N.; Lou, X.; Fortina, P.; Kricka, L.; Wilding, P. Surface effects on PCR reactions in multichip microfluidic platforms. *Biomed. Microdevices* **2004**, *6*, 75–80. [CrossRef]
48. Giordano, B.; Copeland, E.; Landers, J. Towards dynamic coating of glass microchip chambers for amplifying DNA via the polymerase chain reaction. *Electrophoresis* **2001**, *22*, 334–340. [CrossRef]
49. Shoffner, M.; Cheng, J.; Hvichia, G.; Kricka, L.; Wilding, P. Chip PCR. I. Surface passivation of microfabricated silicon-glass chips for PCR. *Nucleic Acids Res.* **1996**, *24*, 375–379. [CrossRef]
50. Felbel, J.; Bieber, I.; Pipper, J.; Köhler, J. Investigations on the compatibility of chemically oxidized silicon (SiO_x)-surfaces for applications towards chip-based polymerase chain reaction. *Chem. Eng. J.* **2004**, *101*, 333–338. [CrossRef]
51. Schott. Schott Borofloat 33. 2017. Available online: https://psec.uchicago.edu/glass/borofloat_33_e.pdf (accessed on 24 June 2020).
52. Stankova, N.; Atanasov, P.; Nikov, R.; Nikov, R.; Nedyalkov, N.; Stoyanchov, T.; Fukata, N.; Kolev, K.; Valova, E.; Georgieva, J.; et al. Optical properties of polydimethylsiloxane (PDMS) during nanosecond laser processing. *App. Surf. Sci.* **2016**, *374*, 96–103. [CrossRef]
53. TOPAS Advanced Polymers. TOPAS COC Polymers. 2017. Available online: https://topas.com/products/topas-coc-polymers (accessed on 26 June 2020).
54. Cui, F.; Chen, W.; Wu, X.; Guo, Z.; Liu, W.; Zhang, W.; Chen, W. Design and experiment of a PDMS-based PCR chip with reusable heater of optimized electrode. *Microsyst. Technol.* **2017**, *23*, 3069–3079. [CrossRef]

55. Glassbrenner, C.; Slack, G. Thermal conductivity of silicon and germanium from 3 K to the melting point. *Phys. Rev.* **1964**, *134*, A1058. [CrossRef]
56. Corning, D. Sylgard 184 Silicone Elastomer. 2014. Available online: http://www.dowcorning.com/DataFiles/090276fe80190b08.pdf (accessed on 26 June 2020).
57. Sato, K.; Shikida, M.; Matsushima, Y.; Yamashiro, T.; Asaumi, K.; Iriye, Y.; Yamamoto, M. Characterization of orientation-dependent etching properties of single-crystal silicon: Effects of KOH concentration. *Sens. Actuators A Phys.* **1998**, *64*, 87–93. [CrossRef]
58. Lee, J.; Park, C.; Whitesides, G. Solvent compatibility of poly (dimethylsiloxane)-based microfluidic devices. *Anal. Chem.* **2003**, *75*, 6544–6554. [CrossRef] [PubMed]
59. McDonald, J.; Duffy, D.; Anderson, J.; Chiu, D.; Wu, H.; Schueller, O.; Whitesides, G. Fabrication of microfluidic systems in poly(dimethylsiloxane). *Electrophoresis* **2000**, *21*, 27–40. [CrossRef]
60. Christensen, T.; Pedersen, C.; Gröndahl, K.; Jensen, T.; Sekulovic, A.; Bang, D.; Wolff, A. PCR biocompatibility of lab-on-a-chip and MEMS materials. *J. Micromech. Microeng.* **2007**, *17*, 1527. [CrossRef]
61. Pekin, D.; Skhiri, Y.; Baret, J.; Le Corre, D.; Mazutis, L.; Salem, C.; Millot, F.; El Harrak, A.; Hutchison, J.; Larson, J.; et al. Quantitative and sensitive detection of rare mutations using droplet-based microfluidics. *Lab Chip* **2011**, *11*, 2156–2166. [CrossRef]
62. Marcus, J.; Anderson, W.; Stephen, R. Parallel picoliter RT-PCR assays using microfluidics. *Anal. Chem.* **2006**, *78*, 956–958. [CrossRef]
63. Hatch, A.; Fisher, J.; Pentoney, S.; Yang, D.; Lee, A. Tunable 3D droplet self-assembly for ultra-high-density digital micro-reactor arrays. *Lab Chip* **2011**, *11*, 2509–2517. [CrossRef]
64. Kiss, M.; Ortoleva-Donnelly, L.; Beer, N.; Warner, J.; Bailey, C.; Colston, B.; Rothberg, J.; Link, D.; Leamon, J. High-Throughput Quantitative PCR in Picoliter Droplets. *Anal. Chem.* **2008**, *80*, 8975–8981. [CrossRef]
65. Hatch, A.; Ray, T.; Lintecum, K.; Youngbull, C. Continuous flow real-time PCR device using multi-channel fluorescence excitation and detection. *Lab Chip* **2014**, *14*, 562–568. [CrossRef]
66. Geng, T.; Novak, R.; Mathies, R. Single-Cell Forensic Short Tandem Repeat Typing within Microfluidic Droplets. *Anal. Chem.* **2014**, *86*, 703–712. [CrossRef]
67. Nakano, M.; Komatsu, J.; Matsuura, S.; Takashima, K.; Katsura, S.; Mizuno, A. Single-molecule PCR using water-in-oil emulsion. *J. Biotechnol.* **2003**, *102*, 117–124. [CrossRef]
68. Moschou, D.; Vourdas, N.; Kokkoris, G.; Papadakis, G.; Parthenios, J.; Chatzandroulis, S.; Tserepi, A. All-plastic, low-power, disposable, continuous-flow PCR chip with integrated microheaters for rapid DNA amplification. *Sens. Actuators B Chem.* **2014**, *199*, 470–478. [CrossRef]
69. Aboud, M.; Gassmann, M.; McCord, B. The development of mini pentameric STR loci for rapid analysis of forensic DNA samples on a microfluidic system. *Electrophoresis* **2010**, *31*, 2672–2679. [CrossRef]
70. Cho, Y.K.; Kim, J.; Lee, Y.; Kim, Y.A.; Namkoong, K.; Lim, H.; Oh, K.; Kim, S.; Han, J.; Park, C.; et al. Clinical evaluation of micro-scale chip-based PCR system for rapid detection of hepatitis B virus. *Biosens. Bioelectron.* **2006**, *21*, 2161–2169. [CrossRef]
71. Di Carlo, D.; Ionescu-Zanetti, C.; Zhang, Y.; Hung, P.; Lee, L. On-chip cell lysis by local hydroxide generation. *Lab Chip* **2004**, *5*, 171–178.
72. Nevill, J.; Cooper, R.; Dueck, M.; Breslauer, D.; Lee, L. Integrated microfluidic cell culture and lysis on a chip. *Lab Chip* **2007**, *7*, 1689–1695. [CrossRef]
73. Jen, C.P.; Hsiao, J.H.; Maslov, N. Single-cell chemical lysis on microfluidic chips with arrays of microwells. *Sensors* **2011**, *12*, 347–358. [CrossRef]
74. Tsougeni, K.; Papadakis, G.; Gianneli, M.; Grammoustianou, A.; Constantoudis, V.; Dupuy, B.; Petrou, P.; Kakabakos, S.; Tserepi, A.; Gizeli, E.; et al. Plasma nanotextured polymeric lab-on-a-chip for highly efficient bacteria capture and lysis. *Lab Chip* **2016**, *16*, 120–131. [CrossRef]
75. Reedy, C.; Price, C.; Sniegowski, J.; Ferrance, J.; Begley, M.; Landers, J. Solid phase extraction of DNA from biological samples in a post-based, high surface area poly (methyl methacrylate)(PMMA) microdevice. *Lab Chip* **2011**, *11*, 1561–1700. [CrossRef]
76. Zhang, X.; Wu, X.; Peng, R.; Li, D. Electromagnetically controlled microfluidic chip for DNA extraction. *Measurement* **2015**, *75*, 23–28. [CrossRef]
77. Obeid, P.; Christopoulos, T.; Crabtree, H.; Backhouse, C. Microfabricated device for DNA and RNA amplification by continuous-flow polymerase chain reaction and reverse transcription-polymerase chain reaction with cycle number selection. *Anal. Chem.* **2003**, *75*, 288–295. [CrossRef] [PubMed]

78. Mitnik, L.; Carey, L.; Burger, R.; Desmarais, S.; Koutny, L.; Wernet, O.; Matsudaira, P.; Ehrlich, D. High-speed analysis of multiplexed short tandem repeats with an electrophoretic microdevice. *Electrophoresis* **2002**, *23*, 719–726. [CrossRef]
79. Obeid, P.; Christopoulos, T. Continuous-flow DNA and RNA amplification chip combined with laser-induced fluorescence detection. *Anal. Chim. Acta* **2003**, *494*, 1–9. [CrossRef]
80. Price, C.; Leslie, D.; Landers, J. Nucleic acid extraction techniques and application to the microchip. *Lab Chip* **2009**, *9*, 2484–2494. [CrossRef] [PubMed]
81. Duarte, G.; Price, C.; Augustine, B.; Carrilho, E.; Landers, J. Dynamic Solid Phase DNA Extraction and PCR Amplification in Polyester-Toner (PeT) Based Microchip. *Anal. Chem.* **2011**, *83*, 5182–5189. [CrossRef]
82. Hagan, K.; Reedy, C.; Bienvenue, J.; Dewald, A.; Landers, J. A valveless microfluidic device for integrated solid phase extraction and polymerase chain reaction for short tandem repeat (STR) analysis. *Analyst* **2011**, *136*, 1928–1937. [CrossRef]
83. Kopp, M.; Mello, A.; Manz, A. Chemical amplification: continuous-flow PCR on a chip. *Science* **1998**, *280*, 1046–1048. [CrossRef]
84. Abgrall, P.; Gue, A. Lab-on-chip technologies: making a microfluidic network and coupling it into a complete microsystem: A review. *J. Micromech. Microeng.* **2007**, *17*, R15. [CrossRef]
85. Frey, O.; Bonneick, S.; Hierlemann, A.; Lichtenberg, J. Autonomous microfluidic multi-channel chip for real-time PCR with integrated liquid handling. *Biomed. Microdevices* **2007**, *9*, 711–718. [CrossRef]
86. Law, I.; Loo, J.; Kwok, H.; Yeung, H.; Leung, C.; Hui, M.; Wu, S.; Chan, H.; Kwan, Y.; Ho, H.; et al. Automated real-time detection of drug-resistant Mycobacterium tuberculosis on a lab-on-a-disc by Recombinase Polymerase Amplification. *Anal. Biochem.* **2018**, *544*, 98–107. [CrossRef]
87. Bhagat, A.; Jothimuthu, P.; Papautsky, I. Photodefinable polydimethylsiloxane (PDMS) for rapid lab-on-a-chip prototyping. *Lab Chip* **2007**, *7*, 1192–1197. [CrossRef]
88. Morganti, E.; Collini, C.; Potrich, C.; Ress, C.; Adami, A.; Lorenzelli, L.; Pederzolli, C. A Micro Polymerase Chain Reaction Module for Integrated and Portable DNA Analysis Systems. *J. Sens.* **2011**, *2011*, 1–7. [CrossRef]
89. Yoon, D.; Lee, Y.S.; Lee, Y.; Cho, H.; Sung, S.; Oh, K.; Cha, J.; Lim, G. Precise temperature control and rapid thermal cycling in a micromachined DNA polymerase chain reaction chip. *J. Micromech. Microeng.* **2002**, *12*, 813. [CrossRef]
90. Hu, G.; Xiang, Q.; Fu, R.; Xu, B.; Venditti, R.; Li, D. Electrokinetically controlled real-time polymerase chain reaction in microchannel using Joule heating effect. *Anal. Chim. Acta* **2006**, *557*, 146–151. [CrossRef]
91. Xiang, Q.; Xu, B.; Fu, R.; Li, D. Real time PCR on disposable PDMS chip with a miniaturized thermal cycler. *Biomed. Microdevices* **2005**, *7*, 273–279. [CrossRef]
92. Kim, J.; Lee, J.; Seong, S.; Cha, S.; Lee, S.; Kim, J.; Park, T. Fabrication and characterization of a PDMS–glass hybrid continuous-flow PCR chip. *Biochem. Eng. J.* **2006**, *29*, 91–97. [CrossRef]
93. Kuo, J.; Chiu, D. Disposable microfluidic substrates: transitioning from the research laboratory into the clinic. *Lab Chip* **2011**, *11*, 2656–2665. [CrossRef]
94. Tsao, C.W.; DeVoe, D. Bonding of thermoplastic polymer microfluidics. *Microfluid. Nanofluidics* **2009**, *6*, 1–16. [CrossRef]
95. Bruijns, B.; Veciana, A.; Tiggelaar, R.; Gardeniers, H. Cyclic Olefin Copolymer Microfluidic Devices for Forensic Applications. *Biosensors* **2019**, *9*, 85. [CrossRef]
96. Bruijns, B.; Costantini, F.; Lovecchio, N.; Tiggelaar, R.; Di Timoteo, G.; Nascetti, A.; de Cesare, G.; Gardeniers, J.; Caputo, D. On-chip real-time monitoring of multiple displacement amplification of DNA. *Sens. Actuators B Chem.* **2019**, *293*, 16–22. [CrossRef]
97. Bruijns, B. Microfluidic Devices for Presumptive Forensic Tests. PhD Thesis, University of Twente, Enschede, The Netherlands, 2019.
98. Ogilvie, I.; Sieben, V.; Floquet, C.; Zmijan, R.; Mowlem, M.; Morgan, H. Reduction of surface roughness for optical quality microfluidic devices in PMMA and COC. *J. Micromech. Microeng.* **2010**, *20*, 065016. [CrossRef]
99. Serra, M.; Pereiro, I.; Yamada, A.; Viovy, J.L.; Descroix, S.; Ferraro, D. A simple and low-cost chip bonding solution for high pressure, high temperature and biological applications. *Lab Chip* **2017**, *17*, 629–634. [CrossRef]

100. GE Healthcare. illustra GneomiPhi V2 DNA Amplification Kit, Product Web Protocol. 2006. Available online: https://www.cytivalifesciences.co.jp/tech_support/manual/pdf/25660030wp.pdf (accessed on 26 June 2020).
101. Dean, F.; Hosono, S.; Fang, L.; Wu, X.; Faruqi, A.; Bray-Ward, P.; Sun, Z.; Zong, Q.; Du, Y.; Du, J.; et al. Comprehensive human genome amplification using multiple displacement amplification. *Proc. Natl. Acad. Sci. USA* **2002**, *99*, 5261. [CrossRef]
102. Kumar, G.; Garnova, E.; Reagin, M.; Vidali, A. Improved multiple displacement amplification with φ29 DNA polymerase for genotyping of single human cells. *Biotechniques* **2008**, *44*, 879–890. [CrossRef]
103. Kim, A.R.; Park, T.; Kim, M.; Kim, I.H.; Kim, K.S.; Chung, K.; Ko, S. Functional fusion proteins and prevention of electrode fouling for a sensitive electrochemical immunosensor. *Anal. Chim. Acta* **2017**, *967*, 70–77. [CrossRef]
104. Lee, T.; Han, K.; Barrett, D.; Park, S.; Soper, S.; Murphy, M. Accurate, predictable, repeatable micro-assembly technology for polymer, microfluidic modules. *Sens. Actuators B Chem.* **2018**, *254*, 1249–1258. [CrossRef]
105. Chin, W.; Sun, Y.; Høgberg, J.; Hung, T.; Wolff, A.; Bang, D. Solid-phase PCR for rapid multiplex detection of Salmonella spp. at the subspecies level, with amplification efficiency comparable to conventional PCR. *Anal. Bioanal. Chem.* **2017**, *409*, 2715–2726. [CrossRef]
106. Kodzius, R.; Xiao, K.; Wu, J.; Yi, X.; Gong, X.; Foulds, I.and Wen, W. Inhibitory effect of common microfluidic materials on PCR outcome. *Sens. Actuators B Chem.* **2012**, *161*, 349–358. [CrossRef]
107. Erill, I.; Campoy, S.; Erill, N.; Barbé, J.; Aguiló, J. Biochemical analysis and optimization of inhibition and adsorption phenomena in glass–silicon PCR-chips. *Sens. Actuators B Chem.* **2003**, *96*, 685–692. [CrossRef]
108. Taylor, T.; Winn-Deen, E.; Picozza, E.; Woudenberg, T.; Albin, M. Optimization of the performance of the polymerase chain reaction in silicon-based microstructures. *Nucleic Acids Res.* **1997**, *25*, 3164–3168. [CrossRef]
109. Kreader, C. Relief of amplification inhibition in PCR with bovine serum albumin or T4 gene 32 protein. *Appl. Environ. Microbiol.* **1996**, *62*, 1102–1106. [CrossRef] [PubMed]
110. Qin, K.; Lv, X.; Xing, Q.; Li, R.; Deng, Y. A BSA coated NOA81 PCR chip for gene amplification. *Anal. Methods* **2016**, *8*, 2584–2591. [CrossRef]
111. Sweryda-Krawiec, B.; Devaraj, H.; Jacob, G.; Hickman, J. A new interpretation of serum albumin surface passivation. *Langmuir* **2004**, *20*, 2054–2056. [CrossRef]
112. Jeyachandran, Y.; Mielczarski, E.; Rai, B.; Mielczarski, J. Quantitative and qualitative evaluation of adsorption/desorption of bovine serum albumin on hydrophilic and hydrophobic surfaces. *Langmuir* **2009**, *25*, 11614–11620. [CrossRef] [PubMed]
113. Rhee, M.; Light, Y.; Yilmaz, S.; Adams, P.; Saxena, D.; Meagher, R.; Singh, A. Pressure stabilizer for reproducible picoinjection in droplet microfluidic systems. *Lab Chip* **2014**, *14*, 4533–4539. [CrossRef]
114. Sidore, A.; Lan, F.; Lim, S.; Abate, A. Enhanced sequencing coverage with digital droplet multiple displacement amplification. *Nucleic Acids Res.* **2015**, *44*, e66. [CrossRef]
115. Leung, K.; Klaus, A.; Lin, B.; Laks, E.; Biele, J.; Lai, D.; Bashashati, A.; Huang, Y.F.; Aniba, R.; Moksa, M.; et al. Robust high-performance nanoliter-volume single-cell multiple displacement amplification on planar substrates. *Proc. Natl. Acad. Sci. USA* **2016**, *113*, 8484–8489. [CrossRef]
116. Kim, S.; Premasekharan, G.; Clark, I.; Gemeda, H.; Paris, P.; Abate, A. Measurement of copy number variation in single cancer cells using rapid-emulsification digital droplet MDA. *Microsyst. Nanoeng.* **2017**, *3*, 17018. [CrossRef]
117. De Bourcy, C.; De Vlaminck, I.; Kanbar, J.; Wang, J.; Gawad, C.; Quake, S. A quantitative comparison of single-cell whole genome amplification methods. *PLoS ONE* **2014**, *9*, e105585. [CrossRef]
118. Andréasson, H.; Gyllensten, U.; Allen, M. Real-time DNA quantification of nuclear and mitochondrial DNA in forensic analysis. *Biotechniques* **2002**, *33*, 402–411. [CrossRef]

© 2020 by the authors. Licensee MDPI, Basel, Switzerland. This article is an open access article distributed under the terms and conditions of the Creative Commons Attribution (CC BY) license (http://creativecommons.org/licenses/by/4.0/).

Review

Electrochemical (Bio)Sensing of Maple Syrup Urine Disease Biomarkers Pointing to Early Diagnosis: A Review

Sophia Karastogianni * and Stella Girousi

School of Chemistry, Faculty of Science, Aristotle University of Thessaloniki, 54124 Thessaloniki, Greece; girousi@chem.auth.gr
* Correspondence: skarastogianni@hotmail.com; Tel.: +30-6940440809

Received: 4 September 2020; Accepted: 8 October 2020; Published: 9 October 2020

Abstract: Metabolic errors are inherited diseases, where genetic defects prevent a metabolic path, ending up in enzyme malfunction. In correspondence to its remaining or plenitude fall of enzymatic potency, there is an amassment of dangerous metabolites near the metabolic bar and/or a dearth of necessary products, inducing a certain disease. These metabolic errors may include deviations such as point mutations, expunctions or interferences, or further complicated genomic disorders. Based on these facts, maple syrup urine disease (MSUD) is a scarce metabolic disease, generated by huge concentrations of branched-chain amino acids (b AAs), i.e., leucine, isoleucine, and valine. In this situation, these large amounts of b AAs provoke abnormalities such as liver failure, neurocognitive dysfunctions, and probably death. To overpass those problems, it is crucial to implement a timely and agile diagnosis at the early stages of life in view of their immutable consequence on neonates. Thus, this review will describe MSUD and b AAs analysis based on electrochemical (bio)sensing.

Keywords: maple syrup urine disease; branched-chain amino acids; electrochemical (bio)sensing

1. Introduction

Amino acids are important in cellular metabolism, as they constitute the architectural parts for protein synthesis and metabolites, supplying the components for farther reactions occurring in living organisms. Amino acids such as aspartic acid (Asp), glutamic acid (Glu), γ-aminobutyric acid (GABA) and taurine (Tau) operate like neurotransmitters, regulating synaptic transference and recollection [1]. Branched-chain amino acids (b AAs) (leucine (Leu), isoleucine (Ile), and valine (Val)) partake in proteinic synthesis and protein catabolism. Phenylalanine (Phe) and tyrosine (Tyr) participate in the formation of trace amines and catecholamines [1].

Meanwhile, most AAs of nutritional significance lies on L-isomers. Natural proteins are entirely formed from L AAs [2,3]. The body cannot use vitamins or minerals in isolation. The enzymes, hormones, body tissues, even bones are constructed from AAs with vitamins and minerals hook-ups. The vitamins and minerals cannot perform this action without free AAs to create the required hook-ups. Therefore, AAs are necessary for vitamins and minerals to accomplish their task rightly. The free AAs are also needed to preserve neurotransmitters, as these are remarkably valuable for osmoregulation of cells and employed as an energy source.

For that matter, human body acquires 20-times more AAs than vitamins and about 4-times more AAs than minerals [4]. Both the D- and L AAs interconvert one to the other over a period, reaching the equilibrium by racemization. During food processing, the L AAs may be racemized to D-isomers. The concentration of L AA can likewise be used as a measure of the nutritive content of the food. Moreover, the chiral amino acids outline is a profitable mean for scanning countless fermentation processes alongside with the detection of bacterial action. The D AAs have been thought as unnatural AAs.

When D AAs are replaced for L AAs in a protein, the protein goes through a switch in structure. This switch can shift the traits and functions of the protein. It is known that D AAs is a common element of bacterial cell wall [5]. The excretion of D AA in physiological fluids is aroused by age, diet, physiological state, and antibiotic therapies. The raised level of D AAs activates nephrotoxicity, growth inhibition, liver damage, fibrosis, and necrosis of kidney cell together with interference in the biosynthesis of certain fundamental neurotransmitters [6]. The white and gray matter of Alzheimer brain incorporates D AA, 1 to 4 times greater than the relevant sector of normal brains [7]. The human beings with renal breakdown have large levels of D AA in urine and serum [8–10].

On the other hand, rare metabolic diseases appear due to genetic abnormalities in the enzymes of the metabolic paths of dietary components. Generally, a scarce disorder in the broad populace is considered when a currency of at most 1 in 20,000 neonates occurs [11]. Presently, 400 million people all over the world are suffering of a scarce disease, about half of them are toddlers and 30% of those patents pass away in reach the antecedent of five years of living, as these diseases are exceptionally threatening in the cradle [12]. The symptoms, prediction, and precise decline of the catabolic paths are utterly distinctive in different metabolic diseases. Considering the aforementioned reasons, the assortment of the above-mentioned maladies is not rigorously prescribed and the most prevalent assortment takes into consideration the dominant molecule influenced (carbohydrates, fatty acids, AA and organic acids), provoking carbohydrate, fatty acid, amino, and organic acids-based diseases, respectively [13,14].

Amino acids maladies displace an autosomal recessive manner of inheritance which suggests that the mutation created a metabolic block is existing in the genetic element of both parents. Because of mutation, the inherited flaw is expressed downstream as a deficiency or a fragmentary biological activity of enzymes engaged in amino acids metabolism. Therefore, some substrates in these paths increase or are switched into different paths. Accordingly, amino acids complications are biochemically outlined by unusual levels of single or several amino acids and their downstream plasma and/or urine metabolites. Amino acid abnormalities are performed with variable and often nonspecific clinical symptoms. In alliance with medical assistance, these disorders are handled by nutritional constraints, supplements, and pharmaceutical food. In this sense, diseases due to amino acid disorders are one of the most dangerous metabolic diseases owed to their deadly repercussions on nurslings.

In particular, the most common diseases due to amino acid disorders are phenylketonuria (PKU), tyrosinemia, homocystenuria, arginase deficiency and maple syrup urine smelling disease (MSUD). All the particular strokes enjoy resemblances in clinical expression, generating mental obstruction in neonates, dementia, or are correlated with syndromes such as Parkinson's disease. They can also induce liver failure, rickets, hepatocarcinoma and death without medication in the early stage of essence [15,16]. Equivalently, a lack of these items can produce hypochondria, depression or albinism [16,17].

The case of MSUD is a scarce metabolic malady alongside a preponderance of 1:200,000 living childbirth [15]. It is provoked by tremendous amounts of the L-branched-chain amino acid (b AAs), such as leucine, isoleucine, and valine (Leu, Ile, and Val, biomarkers) [18]. In aforementioned malady, these excessive elevations of b AAs provoke complications liver illness, neurological impairments, and indeed death. To prevent these problems, it is particularly notable to obtain a timely and rapid diagnosis in the prime periods of life [14,15].

The most common methodologies that are used in the determination of branched-chain amino acids and therefore MSUD, are MS/MS, enzyme activity assays, HPLC, capillary electrophoresis, and genetic testing [19–27]. Notwithstanding, all the practices are time exhausting and, in a few incidents, crave a huge sample volume. A supplementary obstacle relevant to the diagnosis is the matter that for the newborn the monitoring criteria are different among various countries, and indeed vary in the same country, which makes it more troublesome to diagnose MSUD. In addition, there is no end treatment, but dietary constraints are pinpointed and accordingly, repeated monitoring of the target molecule should be performed during the patient's life to evade injurious influences.

Problems related to the diagnosis and monitoring of MSUD or the detection of L b AAs underscore the need for developing new diagnostic methods using easier-to-use, low-sample-volume approaches, which is particularly important in neonates [28–30]. These requirements are met by electrochemical methods and electrochemical (bio)sensors, which are a reliable and profitable appliance for the determination of metabolic biological markers, in order to facilitate their use in the diagnosis and monitoring of this scarce disease, due to their ease of use, simplicity, selectivity, sensibility and low cost [31,32]. In this manner, electrochemical method could be expanded in early diagnosis of other inborn errors of metabolism, including carbohydrate, fatty acid, and amino and organic acids-based diseases [33].

As it can be seen by the lack of literature, electrochemical sensors have been rarely used in the detection of b AAs related to rare diseases such as MSUD clinical diagnostics. Therefore, in this review, electrochemical (bio)sensors for the determination of branched-chain amino acids are summarized.

2. Maple Syrup Urine Disease

Amino acids (Figure 1) are necessary constitutional protein entities and precursors of neurotransmitters, porphyrins, and nitric oxide. Moreover, dietary proteins contain certain amino acids which are catabolized in human body and form organic acids, replenishing Krebs cycle and ammonia that expunge over the urea cycle and accordingly acts as an energy source [34].

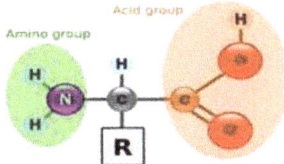

Figure 1. Structure of amino acid.

Aminoacidopathies (amino acids disorders) are a class of hereditary flaws of metabolism strokes, induced by the acquired deformities in paths engaged in amino acid metabolism. Primary amino acids disorders are caused due to mutations, resulting a metabolic block in both parents [34]. Consequently, the flaw is expressed downstream as a shortfall or a limited biological activity of enzymes engaged in amino acids metabolism [34]. Therefore, part of substrates in the particular paths acquire or are whirled toward surrogate paths. In other words, these maladies are biochemically outlined by irregular amounts of single or several amino acids and their metabolites. Aminoacidopathies bear a range of nonspecific clinical symptoms [34]. In alliance to therapeutic alimentation, these afflictions are handled by dietary constraints, supplements, and pharmaceutical foods, reducing the expenditure of a disturbing amino acid or in number incidents, protein expenditure [34,35].

MSUD is an autosomal relapsing disease which is the product of the lack of branched-chain α-keto acid dehydrogenase complex (bkAD) in the second step of catabolic path of b AAs [35]. MSUD is performed alongside five clinical phenotypes without a precise genotype-phenotype relationship. MSUD types can be classified according to the age at onset, the harshness of symptoms, the reply to thiamine supplementation and the biochemical outcomes [35]. In neonatal period classic and E3-deficient MSUD ordinarily appears, while the intermediate, intermittent, and thiamine-responsive types occur in any time of life [35]. It is presented by neurological and developmental delay, encephalopathy, feeding problems, and a maple syrup odor to the urine. It is as well biochemically described by risen plasma b AAs [34,35]. It is regulated by dietary leucine constraints; thus, all b AAs and allo-isoleucine are customarily monitored and alongside prompt medication, inmates generally enjoy satisfying clinical results [34,35].

Biallelic pathogenic variations in the catalytic segments of bkAD reduce its activity with reinforcing b AA amounts and inducing toxicity inward skeletal muscle and brain tissue [36,37]. b AA catabolism is imperative for regular functions [38]. The prime step happens in the mitochondria and includes the alteration of leucine, isoleucine, and valine into their related α-ketoacids by branch-chain aminotransferase. b AAs can be identified in protein-rich nutrition and rest in the family of the nine amino acids indispensable for mankind, presenting influential provinces in protein synthesis and function, cellular signaling, and glucose metabolism [39,40].

On the next step in b AA catabolism, the bkAD complex commence oxidative decarboxylation of α-ketoacids [14], resulting in the alteration of α-ketoacids into acetoacetate, acetyl-CoA, and succinyl-CoA. The bkAD complex is composed of different segments, incorporating subunits E1α and E1β, E2, and E3. Reinforced b AA concentrations in the body renounce in these ingredients inducing MSUD [14,37,38]. Mouse models were used [41] as well as MSUD patients [42], where exaggerated loads of b AAs emerged and can generate serious tissue corruption if omitted without treatment. Abnormalities in b AA metabolism may induce implementations in glutamate synthesis, causing neurological implications [35]. The solution in preventing this symptomatology is the control of plasma concentrations of b AAs. In addition, the accretion of Leu is extremely neurotoxic [37] and excessive concentration of Leu may influence water homeostasis into the subcortical gray matter, resulting in bloating within the brain, convert nitrogen homeostasis farther draining glutamate amounts, reinforce oxidative stress, and wrestle along other amino acids, such as Tyr, in the central nervous system (CNS), which is engaged in protein signaling [35]. Moreover, α-ketoisocaproic acid, which is regarded to be an intervening in the metabolism of Leu, is a neurotoxin, advancing the encephalopathic disorder [35].

3. Conventional Detection Methods of Amino Acids

Different analytical techniques, such as high performance liquid chromatography (HPLC) and capillary electrophoresis (CE), have been adopted in the detection of AAs or b AAs. In recent past, ongoing types of columns, such as sub-2 μm-particle packed, monolithic silica, and core-shell columns, have been broadly employed in liquid chromatography (LC) testing of AA and the analysis time has been largely minimized. The on-chip LC methodology was also established for AA detection. Separation methods have been conjoined with varied determination processes, incorporating UV [23,24], FL [21,43,44], MS [45–47], and electrochemical [25,26,48] detection. Furthermore, MS has been broadly used and has turned out to be the most accepted determination method in AAs analysis. In summary, Song et al. [1] newly reviewed the recent trends in conventional analytical methods of amino acids in biological samples, where detailed information is given about their analytical features and advantages.

4. Electrochemical bAA Sensors and Biosensors

Electrochemical techniques rest on the fabrication of sensors or biosensors and are considered suitable for in situ determination of substances because they are extremely sensitive, simple, reproducible, cheep, relatively fast, and direct, without the use of extraction or preconcentration steps [49,50]. They can also be easily used in miniaturization [51]. Electrochemical detection is usually based on monitoring the signal of oxidation or reduction of the electroactive groups accumulated on the electrode surface [51] and measure features such as electrode potential, current intensity, the amount of electricity passing through the cell, resistance and the time [52].

Cyclic voltammetry (CV) is commonly used on electrochemical procedures, due to its qualitative and quantitative information. By applying CV, a wide range of sensors and biosensors could be developed, determining the analyte. Differential pulse voltammetry (DPV) and square-wave voltammetry (SWV) are pulsed methods [53,54]. The excitation potential in SWV inheres of a symmetrical square-wave pulse of a settled amplitude superimposed on a staircase waveform of step height. In this method the forward pulse of the square wave concurs with the staircase step [55]. DPV is a voltammetric method, comparable to SWV, where the potential excitation inheres of small pulses, which are superimposed upon a staircase waveform [56]. The major profit of the particular methods is the insignificant capacitive current which emerges on the enrichment of the sensitivity of the pulse voltammetric strategies. However, DPV is mostly administered on irreversible processes or on systems posing slow-reaction kinetics, whereas SWV is usually administered on the examination of reversible processes (rapid reaction kinetics systems) [56,57]. Electrochemical sensors have been hardly practiced in the detection of b AAs such as Leu, Ile, and Val, associated with rare diseases clinical diagnostics. This is expounded by the evidence that these substances are electrochemically inert on bare electrodes [33]. Thus, in the subsequence chapters, electrochemical (bio)sensors for the determination of b AA are summarized.

4.1. Metal Nanoparticles and b AA Electrochemical (Bio)Sensing

The limited use of electrochemical sensors in the detection of b AA could be explained by the fact that these molecules are electrochemically inactive on bare electrodes. This limitation is can be solved by the modification of electrode's surface using metal nanoparticles. On this ground, iron oxides [58,59] CoNPs [60], strontium nanorods [61] as well as multiwall carbon nanotubes (MWCNTs) [62] have been used on the detection of some branched-chain amino acids.

Meanwhile, metal nanoparticles (NPs), such as silver and gold, have exceptional attributes, such as biocompatibility, high conductivity and high surface-to-volume ratio [63]. Thus, they are particularly alluring components for applications in electrochemical sensing and biosensing [64–67], as they are widely used in shaping the surface of electrodes in order to develop methodologies for detecting species of biological interest or to make diagnostic tools for various pathological conditions. Compared to stabilizers such as plant leaves [68], fruit extracts [69], plant roots [70], glucose [71] and carbonates materials [72], used in their composition, organic and natural dyes impart to the formed nanoparticles improved properties [73–75]. In addition, organic dyes have an advantage over the above factors because they have specific ionic, polar, non-bond functional groups (-azo dyes, -sulfites, -hydroxyl, and -nitro groups) and are usually systems that have π-conjugates [76] capable of being polymerized.

For example, Hasanzadeh et al. [59] used the magnetic (Fe_2O_3) mobile crystalline material-41 (MCM-41) to modify the surface of glassy carbon electrode (GCE). They found that the proposed electrode owned electrocatalytic activity against the electro-oxidation of the studied amino acids. In this work the amino acids at larger concentrations were determined by CV, hydrodynamic amperometry, and flow injection analysis. The linear range of the proposed method was in the range of 97–176 nM and the detection limit was found to be equal to 94 nM in the case of Val. The proposed sensor was shown to have rapid response, great catalytic activity, and ease of preparation.

Furthermore, Saghatforoush et al. [60] immobilized a Fe (III)-Schiff base on a modified GCE with multiwall carbon nanotubes (MWCNTs). They discovered that the proposed electrodes had great catalytic activity against the oxidation of amino acids at positive potential in acidic solution. The outcomes gave confirmation that these electrodes postured innate stability at a wide pH range, agile response, great sensitivity, low detection limit and a very positive oxidation potential of amino acids that declined the influence of interferences of the detection method. The linear concentration range of Val, the detection limits of Val (LOD), the limits quantization of Val (LOQ) and relative standard deviation of the above-mentioned sensor were found to be 25–1000 µM, 1.67 µM, 2.79–27.14 and 2.82%, respectively.

Meanwhile, cobalt hydroxide nanoparticles were practiced on the modification of a GCE (CHM-GC) and were employed on the investigation of the electrochemical behavior of some amino acids by Hasanzadeh et al. [61]. CV, chronoamperometry methods, and steady-state polarization measurements were used on the investigation of the oxidation and its kinetics. The results exposed that cobalt hydroxide sponsors the rate of oxidation by reinforcing the peak current, so the particular bimolecular reactions are oxidized at smaller potentials. CVs and chronoamperometry revealed a catalytic EC mechanism to be employable with the electrogeneration of Co(IV).

Strontium oxide nanorods (SrO NR) is another example of metal nanoparticle that can be used in the detection assays of b AAs. Thus, Hussain et al. [62] synthesized SrO NR in alkaline medium by a wet-chemical method. Their results showed that a thin-layer of the NR was accumulated on a GCE, fabricating an electrochemical sensor for L-Leu. The proposed sensor had good sensitivity, a wide dynamic range, and good long-term stability. The response to Leu was studied by the current-voltage (I-V) technique. The calibration plot was linear between 0.1 nM to 0.1 mM. The sensitivity was equal to 2.53 nA·µM^{-1}·cm^{-2}, and the limit of detection was calculated and found equal to 37.5 ± 0.2 pM. The sensor was applied to real samples such as L-Leu spiked urine, milk, and serum, giving acceptable results.

Another useful tool on b AA analysis is multiwall carbon nanotubes. On this matter, Rezaei and Zare [63] developed a simple and sensitive leucine voltammetric detection assay in blood and urine samples. In their study a GCE was used and modified with MWNTs. The CV measurements revealed that MWNTs enhanced the oxidation of Leu GCE. They revealed that Leu was oxidized following multistep mechanism on the proposed electrode. A calibration curve was plotted under the optimum condition, and the sensor had a linear response in the range 9.0×10^{-6}– 1.5×10^{-3} mol L^{-1}. The LOD was found equal to 3.0×10^{-6} mol L^{-1} and a relative standard deviation (RSD%) was estimated below 3.0% (n = 5).

The use of metal nanowires (NWs) gives some advantages in clinical diagnosis such as simplicity, rapid sensor response and short total analysis time, and low sample consumption. Recently, García-Carmona et al. [77] demonstrated the fabrication of vertically aligned nickel nanowires-based electrochemical sensors (v-NiNWs) for fast determination of b AA, aiming to noninvasive screening of MSUD. v-NiNWs. The analytical features of the proposed methodology (for Leu as representative b AA in MSUD) such as LOD (8 mM) and linear range (25–700 mM) demonstrate that v-NiNWs are suitable disposable features for monitoring b AA in MSUD due to their ability to differentiate healthy and MSUD penitents. In addition, the results pointed an excellent intra and inter-electrodes repeatability. Total b AAs were also determined in positive samples with accuracy in just 5 min and using only 250 mL.

Furthermore, NiO NPs were electrochemically immobilized on a GCE and a platinum electrode [78]. In the particular study, CV in a flow cell was used, evaluating the sensors' capability to detect Val among other amino acid. The LOD was estimated to be 4 mM. In this work it was found that Val was electroactive sporadically.

4.2. Enzymatic Aproaches and b AA Electrochemical Biosensing

Over the last few decades enzyme biosensors have been widely developed, and it is evident that they are innovative assays in qualitative, as well in quantitative analysis of numerous analytes [79]. Electrochemical enzyme biosensors have distinguished advantages, because they are highly sensitive and specific, portable, cheap, and can be miniaturized and in this way can be used in the point-of-care diagnostic, which make them alluring for clinical analysis and routine measurements [79].

Generally, a biosensor is a device that operates to analyze a sample in the presence of a specific target analyze. Customarily, a biosensor is manufactured from a biological unit, which is roared as molecular recognition element, and a detector based on physicochemical process or transducer [79]. The biological unit (recognition element) is accumulated on the transducer's surface, interacts with target analyte [59]. The variations are then noticed by the transducer, and transformed to measurable signals, used to detect the concentration of the target molecule [79]. Biosensors are divided depending on either the recognition element, such as nucleic acids, antibodies, enzymes and cells, or by the class of the transducer (optical, mass-based, electrochemical, piezoelectric) [79].

On the other hand, electrochemical biosensors rely on the electrochemical properties of transducers and substances to be analyzed. This kind of biosensors were established as a development of the first glucose enzyme biosensor [80]. Intrinsically, variation in physicochemical features of elements such as current, voltage, resistance, or superficial charge, risen by oxidation-reduction processes are the output signals. The most popular class of transducers are amperometry, potentiometry, conductometry, and impedimetry.

Therefore, a recent example of an enzyme-based biosensor is that García-Carmona et al. [81] developed. The proposed biosensor is claimed to be fast, simple, selective, and sensible. The proposed biosensor was fabricated with the on-line coupling of millimeter size motors (m-motors) and chronoamperometry for real-time analysis of b AAs in MSUD samples. Thus, an integrated cell device, including reservoirs for motors movement and electrochemical determination, was constructed. L-amino acid oxidase was introduced in m-motors by capillarity without coating chemistry, which is necessary for the enantioselective identification, and afterwards the released H_2O_2 was electrochemically screened with copper microwires. The results showed that the innate m-motor *"self-micro mixing"* characteristics and the extended delivery of new and free enzyme had the advantage on the one hand of avoiding the agitation or the physical adsorption of the enzyme, as well as its chemical bonding on transducer's surface and on any other way it reduced the investigation time.

Stefan-van Staden et al. [82] proposed four amperometric diamond paste-based biosensors and then used in the detection of are for the enantiopurity of Leu [82]. Biosensors' layout used physical accumulation of L- and D-oxidases on the proposed electrodes. In this paper, the characteristics of the different proposed sensors were examined in contrast. The results showed that the sensors were linear in pmol/L to nmol/L level. It was also found that the proposed sensor was reliable in detecting the enantiopurity using Leu as a raw material.

Moreover, Labroo and Cui [83] reported the fabrication of an amperometric bienzyme biosensor based on screen-printed electrodes and used it in the detection of Leu. The proposed biosensor was constructed by immobilizing p-hydroxybenzoate hydroxylase (HBH) and Leu dehydrogenase (LDH) on the proposed electrode, using $NADP^+$ and p-hydroxybenzoate as cofactors. The operating principle of this biosensor relied on the catalytic ability of LDH towards the specific dehydrogenation of Leu. The resulted NADPH prompts the hydroxylation of p-hydroxybenzoate by HBH in the existence of oxygen to generating 3,4-dihydroxybenzoate, developing an alteration in electron density on the proposed electrode. They claim that this sensor was linear in the range 10–600 µM and LOD was estimated to equal to 2 µM. Conclusively, they found that the sensor was rapid and reproducible with a total analysis time of 5–10 s.

Finally, the fabrication and analytical effectiveness of a bienzyme biosensors for the selective detection of AAs enantiomers using amperometric transduction is described in the literature [84]. The studied enzymes of the proposed method, as well as the mediator (L-Amino acid oxidase, horseradish peroxidase, ferrocene, respectively) was accumulated on a graphite-Teflon electrode with physical insertion in a graphite-Teflon solution. Experiments were made with and without the regeneration of electrode's surface on the useful lifetime of one single biosensor and on the reproducibility in the fabrication of different biosensors and gave evidence that the constructed biosensor, was robust and reproducible. The proposed modified electrode was employed in the successful detection of enantiomers of AAs in racemic mixtures which was indicative of the selectivity of the method, and to the evaluation of AAs in muscatel grapes.

4.3. Conducting Polymers and b AA Electrochemical (Bio)Sensing

Conducting polymers (CP) are of interest for their use as sensitive electrode surface coatings on electrochemical sensors and biosensors (electrode surface modifiers) [85]. They are outlined by large electrical conductivity and satisfying electrochemical reversibility and therefore upholding their application on sensor transducer signaling. Furthermore, CPs can be chemically acquired functional groups, which act as "tags" because of their qualification to identify biological or chemical items [86,87].

Nevertheless, the determination of small (bioactive) analytes such as b AAs remains an open continues to be an unclosed issue, as their mere correlation with detectable groups on the CP substrate is not adequate to generate the necessary electrochemical change for their detection. The key in this case is to build high-specific CP recognition points, which will strengthen selectivity and advance the sensitivity of the nidification procedure. In connection with the specificity of the above process, molecularly imprinted polymers (MIPs) can be administered in the synthesis of polymers with predetermined molecular recognition features and can be used in constructing sensors and biosensors [88,89]. Molecular imprinting is the innovation of these designs, as they enjoy plentiful improved traits: they are sensible, fast, simple and can be portable [90,91].

Meanwhile, cellulose nanocrystals polymers, usually are used to upgrade sensitivity and selectivity of b AAs detection and may be an alternative solution [92]. This strategy counts on a rod-like network of eminently crystalline fibers and owns a huge specific surface area, contributing valuable electrical and optical attributes. Hence, Bi et al. [90] proposed an electrochemical sensor depended on 2,2,6,6-tetramethylpiperidine-1-oxyl (TEMPO)-oxidized cellulose nanocrystals (TOCNCs). In this study, L-Cys modified Au electrode (TOCNC/L-Cys/Au) was formulated for determination and differentiation of the enantiomers of Phe, Leu, as well as Val. CV and DPV experiments showed that the constructed electrode had a peak current difference for the selected enantiomers. The proposed modifier was outlined by its various interactions with the different enantiomers, obtaining the identification of the enantiomers.

On the other hand, microfluidic chips (MCs) provide many advantages in diagnostics due to theirs fast response, the use of small quantities of sample with good reproducibility, opening new boundaries on point-of-care diagnostics (POC). In addition, electrochemical transducers are convenient for these devices because of their inherent miniaturization and sensitivity. In this sense, conjugated MCs with electrochemical methods which are simple, rapid, and cost-effective, is an attractive approach for a POC device that can be used in metabolic diseases such as MSUD [30,93].

Based on these, an electrochemical microfluidic assay was practiced in the partition, as well as the determination of AAs enantiomers of D-Met and D-Leu by Batalla et al. [94]. The proposed device admitted the adjusted microfluidic D AA partition, as well as the particular reaction among D-amino acid oxidase (DAAO) and one by one AA on a sole arrangement of a MC. The proposed system was claimed to be consistent with small sample expenditure, averting the need for supplements for the partition of enantiomers, and the covalent accumulation of the enzyme on the wall channels or on the electrode surface. Hybrid polymer/graphene-based electrodes were end-channel capped to the microfluidic apparatus, advancing the traits of the procedure. D-Leu were fortuitously determined, adopting the introduced system. D-Leu was detected indirectly by the detection of H_2O_2. The proposed method was claimed to have superlative precision in migration times and in peak heights, revealing the satisfying stability of the proposed sensor. It was found that the proposed sensor had also selectivity, due to the inactiveness of L AAs.

5. Conclusions

In this review, electrochemical (bio)sensors for the detection of b AAs were summarized. Electrochemical (bio)sensors as well as POC devices are relatively new avenues for reliable, accurate, sensitive, selective, green, cheap on the detection of b AA, involved in metabolic diseases such as MSUD, and are in line with current European and worldwide developments, concerning public health issues and representing the state of art on developing analytical methodologies. However, the use of electrochemical (bio)sensors in the detection of b AA have been poorly studied, as it can be seen from the lack of relevant studies in the literature. Although years of research have provided a lot of information on conventional analytical methods such as MS, limited research was made in the field of electrochemical (bio)sensors, which could improve the analytical features of b AA detection. Furthermore, investigating the prospects of increasing the accuracy, the sensitivity, the selectivity, the simplicity as well as minimizing the cost and toxicity of present b AA analytical methods is also an innovative approach to an old need for the world clinical diagnostics and electrochemical (bio)sensor are suitable tools towards this direction. In Table 1 selected studies in electrochemical (bio)sensing MSUD and b AA are summarized.

Table 1. Comparison of selected b AA electrochemical (bio)sensors.

Electrode	Analyte	Linear Range	LOD	Ref
MCM-41-Fe_2O_3/GCE	Valine	97–176 nM	94 nM	[59]
Fe (III)–Schiff base complex on GCE/MWCNTs	Valine	25–1000 µM	1.67 µM	[60]
SrO NR/GCE	L-Leukine	0.1–0.1 mM	37.5 pM	[62]
GCE/MWCNTs	Leukine	9.0×10^{-6}–1.5×10^{-3} M	3×10^{-6} M	[63]
v-NiNWs	Leukine	25–700 mM	8 mM	[77]
p-hydroxybenzoate hydroxylase and leucine dehydrogenase on a screen-printed electrode	Leukine	10 and 600 µM	2 µM	[83]

Abbreviation: GCE: glassy carbon electrode; MCM-41-Fe_2O_3/GCE: modified glassy carbon electrode with magnetic (Fe_2O_3) mobile crystalline material-41 (MCM-41); MWCNTs: multiwall carbon nanotubes; SrO NR/GCE: modified glassy carbon electrode with strontium oxide nanorods; v-NiNWs: vertically aligned nickel nanowires; LOD: limit of detection.

Meanwhile, greater analytical features are attained when electrochemical techniques are coupled with NPs. To that end, the great antifouling trait of NP electrodes is notably meaningful, considering that they are skilled to execute a considerable number of detections without the loss of their analytical features as has been disclosed by their valuable repeatability. This particular trait awards them respectable facilities to be practiced on the determination of biomarkers in real samples.

Notwithstanding, the dominant challenge continues existing, when real samples are to be analyzed, due to problems related to reproducibility, stability, as well as interferences. These elements can be worked out by evolving innovative sensors built on chiral nanostructured components, favoring the selectivity of the determination. Admirable analytical enforcement may be brought about through the coupling of enzymes to NPs and electroactive arbiters.

Conclusively, the captious prospects of the leading edge on the determination of b AAs, clearly launches innovative frontiers on electrochemical sensors for rapid monitoring of a disease, initiating modern concepts in diagnostics. A forthcoming advancement in electrochemical sensing may be the growth of implantable sensors for prolonged disease screening. Thus, novel (bio)materials must integrate into devices, achieving stability and limiting the infections with unwanted substances. Late determination methodologies, such as ultra-fast CV may be occupied for real-time b AA monitoring. The advancement of mercantile arrangements on b AA monitoring predicated on electrochemical (bio)sensors and chromatographic or electrophoretic techniques may be the next step in the field.

Author Contributions: S.K.; investigation, S.K.; writing—original draft preparation, S.K.; writing—review and editing, S.K.; visualization, S.K.; funding acquisition, S.K.; review S.G.; supervision. All authors have read and agreed to the published version of the manuscript.

Funding: This research is co-financed by Greece and the European Union (European Social Fund—ESF) through the Operational Programme «Human Resources Development, Education and Lifelong Learning» in the context of the project "Reinforcement of Postdoctoral Researchers—2nd Cycle" (MIS-5033021), implemented by the State Scholarships Foundation (IKY), grant number MIS 5033021.

Conflicts of Interest: The authors declare no conflict of interest and the funders had no role in the design of the study; in the writing of the manuscript, or in the decision to publish this review.

References

1. Song, Y.; Xu, C.; Kuroki, H.; Liao, Y.; Tsunoda, M. Recent trends in analytical methods for the determination of amino acids in biological samples. *J. Pharm. Biomed. Anal.* **2018**, *147*, 35–49. [CrossRef] [PubMed]
2. Meister, A. Intermediary metabolism of the amino acids. Biochem. *Amino Acids* **1965**, 593–1020. [CrossRef]
3. Marchelli, R. The potential of enantioselective analysis as a quality control tool. *Trends Food Sci. Technol.* **1996**, *7*, 113–119. [CrossRef]
4. Kwan, R.C.; Hon, P.Y.; Renneberg, R. Amperometric biosensor for rapid determination of alanine. *Anal. Chim. Acta* **2004**, *523*, 81–88. [CrossRef]

5. Khoronenkova, S.V.; Tishkov, V.I. D-Amino acid oxidase: Physiological role and applications. *Biochemistry* **2008**, *73*, 1511–1518. [CrossRef] [PubMed]
6. Friedman, M. Origin, microbiology, nutrition, and pharmacology of D-amino acids. *Chem. Biodivers.* **2010**, *7*, 1491–1530. [CrossRef]
7. Kasai, H.; Fukuda, M.; Watanabe, S.; Hayashi-Takagi, A.; Noguchi, J. Structural dynamics of dendritic spines in memory and cognition. *Trends Neurosci.* **2010**, *33*, 121–129. [CrossRef]
8. Kawazoe, T.; Tsuge, H.; Pilone, M.S.; Fukui, K. Crystal structure of human Damino acid oxidase: Context-dependent variability of the backbone conformation of the VAAGL hydrophobic stretch located at the si-face of the flavin ring. *Protein Sci.* **2006**, *15*, 2708–2717. [CrossRef]
9. Bruckner, H.; Hausch, M. D-amino acids in dairy products: Detection, origin and nutritional aspects. I. Milk, fermented milk, fresh cheese and acid curd cheese. *Milchwissenschaft* **1990**, *45*, 357–360.
10. Bruckner, H.; Schieber, A. Determination of free D-amino acids in mammalian by chiral gas chromatography–mass spectrometry. *J. High Resol. Chromatogr.* **2000**, *23*, 576–582. [CrossRef]
11. Pavan, S.; Rommel, K.; Marquina, M.E.M.; Hohn, S.; Lanneau, V.; Rath, A. Clinical Practice Guidelines for Rare Diseases: The Orphanet Database. *PLoS ONE* **2017**, *12*, e0170365. [CrossRef] [PubMed]
12. Piras, D.; Locci, E.; Palmas, F.; Ferino, G.; Fanos, V.; Noto, A.; D'aloja, E.; Finco, G. Rare disease: A focus on metabolomics. *Expert Opin. Orphan Drugs* **2016**, *4*, 1229–1237. [CrossRef]
13. Jumbo-Lucioni, P.P.; Garber, K.; Kiel, J.; Baric, I.; Berry, G.T.; Bosch, A.; Burlina, A.; Chiesa, A.; Pico, M.L.C.; Estrada, S.C.; et al. Diversity of approaches to classic galactosemia around the world: A comparison of diagnosis, intervention, and outcomes. *J. Inherit. Metab. Dis.* **2012**, *35*, 1037–1049. [CrossRef]
14. Burrage, L.C.; Nagamani, S.C.; Campeau, P.M.; Lee, B.H. Branched-chain amino acid metabolism: From rare Mendelian diseases to more common disorders. *Hum. Mol. Genet.* **2014**, *23*, R1–R8. [CrossRef] [PubMed]
15. Scott, C.R. The genetic tyrosinemias. *Am. J. Med. Genet. Part C Semin. Med. Genet.* **2006**, *142C*, 121–126. [CrossRef] [PubMed]
16. García-Cazorla, A.; Wolf, N.; Serrano, M.; Moog, U.; Perez-Duenas, B.; Poo, P.; Pineda, M.; Campistol, J.; Hoffmann, G. Mental retardation and inborn errors of metabolism. *J. Inherit. Metab. Dis.* **2009**, *32*, 597–608. [CrossRef] [PubMed]
17. Harms, E.; Olgemöller, B. Neonatal Screening for Metabolic and Endocrine Disorders. *Dtsch. Arztebl. Int.* **2011**, *108*, 11–22. [CrossRef]
18. Fujimoto, A.; Okano, Y.; Miyagi, T.; Isshiki, G.; Oura, T. Quantitative Beutler Test for Newborn Mass Screening of Galactosemia Using a Fluorometric Microplate Reader T. *Clin. Chem.* **2000**, *46*, 806–810. [CrossRef] [PubMed]
19. Avilov, V.; Zeng, Q.; Shippy, S.A. Threads for tear film collection and support in quantitative amino acid analysis. *Anal. Bioanal. Chem.* **2016**, *408*, 5309–5317. [CrossRef]
20. Borowczyk, K.; Chwatko, G.; Kubalczyk, P.; Jakubowski, H.; Kubalska, J.; Glowacki, R. Simultaneous determination of methionine and homocysteine by on-column derivatization with o-phtaldialdehyde. *Talanta* **2016**, *161*, 917–924. [CrossRef]
21. Azuma, K.; Hirao, Y.; Hayakawa, Y.; Murahata, Y.; Osaki, T.; Tsuka, T.; Imagawa, T.; Okamoto, Y.; Ito, N. Application of pre-column labeling liquid chromatography for canine plasma-free amino acid analysis. *Metabolites* **2016**, *6*, 3. [CrossRef] [PubMed]
22. Jeong, J.; Yoon, H.; Hong, S. Development of a new diagnostic method for galactosemia by high-performance anion-exchange chromatography with pulsed amperometric detection. *J. Chromatogr. A* **2007**, *1140*, 157–162. [CrossRef] [PubMed]
23. Acquaviva, A.; Romero, L.M.; Castells, C.B. Analysis of citrulline and metabolic related amino acids in plasma by derivatization and RPLC. Application of the extrapolative internal standard calibration method. *Microchem. J.* **2016**, *129*, 29–35. [CrossRef]
24. Castellanos, M.; van Eendenburg, C.V.; Gubern, C.; Sanchez, J.M. Ethyl-bridged hybrid column as an efficient alternative for HPLC analysis of plasma amino acids by pre-column derivatization with 6-aminoquinolyl-N-hydroxysuccinimidyl carbamate. *J. Chromatogr. B* **2016**, *1029*, 137–144. [CrossRef] [PubMed]
25. Tuma, P.; Gojda, J. Rapid determination of branched chain amino acids in human blood plasma by pressure-assisted capillary electrophoresis with contactless conductivity detection. *Electrophoresis* **2015**, *36*, 1969–1975. [CrossRef] [PubMed]

26. Ulusoy, S.; Ulusoy, H.I.; Pleissner, D.; Eriksen, N.T. Nitrosation and analysis of amino acid derivatives by isocratic HPLC. *RSC Adv.* **2016**, *6*, 13120–13128. [CrossRef]
27. Blau, N.; Shen, N.; Carducci, C. Molecular genetics and diagnosis of phenylketonuria: State of the art. *Expert Rev. Mol. Diagn.* **2014**, *14*, 655–671. [CrossRef]
28. Dincer, C.; Bruch, R.; Kling, A.; Dittrich, P.S.; Urban, G.A. Multiplexed Point-of-Care Testing—xPOCT. *Trends Biotechnol.* **2017**, *35*, 728–742. [CrossRef]
29. Luppa, P.B.; Bietenbeck, A.; Beaudoin, C.; Giannetti, A. Clinically relevant analytical techniques, organizational concepts for application and future perspectives of point-of-care testing. *Biotechnol. Adv.* **2016**, *34*, 139–160. [CrossRef]
30. Zhang, W.; Guo, S.; Carvalho, W.S.P.; Jiang, Y.; Serpe, M.J. Portable point-of-care diagnostic devices. *Anal. Methods* **2016**, *8*, 7847–7867. [CrossRef]
31. Lv, J.; Li, C.; Feng, S.; Chen, S.-M.; Ding, Y.; Chen, C.; Hao, Q.; Yang, T.-H.; Lei, W. A novel electrochemical sensor for uric acid detection based on PCN/MWCNT. *Ionics* **2019**, *25*, 4437–4445. [CrossRef]
32. Zhang, W.; Zhang, X.; Zhang, L.; Chen, G. Fabrication of carbon nanotube-nickel nanoparticle hybrid paste electrodes for electrochemical sensing of carbohydrates. *Sens. Actuators B Chem.* **2014**, *192*, 459–466. [CrossRef]
33. Sandlers, Y. The future perspective: Metabolomics in laboratory medicine for inborn errors of metabolism. *Trans. Res.* **2017**, *189*, 65–75. [CrossRef]
34. García-Carmona, L.; González, M.C.; Escarpa, A. Nanomaterial-based electrochemical (bio)-sensing: One step ahead in diagnostic and monitoring of metabolic rare diseases. *TrAC Trends Anal. Chem.* **2019**, *118*, 29–42. [CrossRef]
35. Blackburn, P.R.; Gass, J.M.; Vairo, F.P.E.; Farnham, K.M.; Atwal, H.K.; Macklin, S.; Klee, E.W.; Atwal, P.S. Maple syrup urine disease: Mechanisms and management. *Appl. Clin. Genet.* **2017**, *10*, 57–66. [CrossRef] [PubMed]
36. Lang, C.H.; Lynch, C.J.; Vary, T.C. BCATm deficiency ameliorates endotoxin-induced decrease in muscle protein synthesis and improves survival in septic mice. *Am. J. Physiol. Regul. Integr. Comp. Physiol.* **2010**, *299*, R935–R944. [CrossRef]
37. Yudkoff, M.; Daikhin, Y.; Nissim, I.; Horyn, O.; Luhovyy, B.; Lazarow, A. Brain amino acid requirements and toxicity: The example of leucine. *J. Nutr.* **2005**, *135*, 1531S–1538S. [CrossRef]
38. Lynch, C.J.; Adams, S.H. Branched-chain amino acids in metabolic signalling and insulin resistance. *Nat. Rev. Endocrinol.* **2014**, *10*, 723–736. [CrossRef]
39. Brosnan, J.T.; Brosnan, M.E. Branched-chain amino acids: Enzyme and substrate regulation. *J. Nutr.* **2006**, *136*, 207S–211S. [CrossRef]
40. Harper, A.E.; Miller, R.H.; Block, K.P. Branched-chain amino acid metabolism. *Annu. Rev. Nutr.* **1984**, *4*, 409–454. [CrossRef]
41. Vogel, K.R.; Arning, E.; Wasek, B.L.; McPherson, S.; Bottiglieri, T.; Gibson, K.M. Brain-blood amino acid correlates following protein restriction in murine maple syrup urine disease. *Orphanet J. Rare Dis.* **2014**, *9*, 73. [CrossRef] [PubMed]
42. Zinnanti, W.J.; Lazovic, J. Interrupting the mechanisms of brain injury in a model of maple syrup urine disease encephalopathy. *J. Inherit. Metab. Dis.* **2012**, *35*, 71–79. [CrossRef] [PubMed]
43. Song, Y.; Takatsuki, K.; Isokawa, M.; Sekiguchi, T.; Mizuno, J.; Funatsu, T.; Shoji, S.; Tsunoda, M. Fast and quantitative analysis of branched-chain amino acids in biological samples using a pillar array column. *Anal. Bioanal. Chem.* **2013**, *405*, 7993–7999. [CrossRef] [PubMed]
44. Sharma, G.; Attri, S.V.; Behra, B.; Bhisikar, S.; Kumar, P.; Tageja, M.; Sharda, S.; Singhi, P.; Singhi, S. Analysis of 26 amino acids in human plasma by HPLC using AQC as derivatizing agent and its application in metabolic laboratory. *Amino Acids* **2014**, *46*, 1253–1263. [CrossRef] [PubMed]
45. Lorenzo, M.P.; Navarrete, A.; Balderas, C.; Garcia, A. Optimization and validation of a CE-LIF method for amino acid determination in biological samples. *J. Pharm. Biomed. Anal.* **2013**, *73*, 116–124. [CrossRef]
46. Delgado-Povedano, M.M.; Calderon-Santiago, M.; Priego-Capote, F.; Luque de Castro, M.D. Study of sample preparation for quantitative analysis of amino acids in human sweat by liquid chromatography-tandem mass spectrometry. *Talanta* **2016**, *146*, 310–317. [CrossRef]

47. Yin, B.; Li, T.; Zhang, S.; Li, Z.; He, P. Sensitive analysis of 33 free amino acids in serum, milk, and muscle by ultra-high-performance liquid chromatography-quadrupole-orbitrap high resolution mass spectrometry. *Food Anal. Methods* **2016**, *9*, 2814–2823. [CrossRef]
48. Tuma, P.; Sustkova-Fiserova, M.; Opekar, F.; Pavlicek, V.; Malkova, K. Large-volume sample stacking for in vivo monitoring of trace levels of gamma-aminobutyric acid, glycine and glutamate in micro dialysates of periaqueductal gray matter by capillary electrophoresis with contactless conductivity detection. *J. Chromatogr. A* **2013**, *1303*, 94–99. [CrossRef]
49. Liu, Y.; Liang, Y.; Yang, R.; Li, J.; Qu, L. A highly sensitive and selective electrochemical sensor based on polydopamine functionalized graphene and molecularly imprinted polymer for the 2,4-dichlorophenol recognition and detection. *Talanta* **2019**, *195*, 691–698. [CrossRef]
50. Ciriello, R.; De Gennaro, F.; Frascaro, S.; Guerrieri, A. A novel approach for the selective analysis of L-lysine in untreated human serum by a co-crosslinked L-lysine–α-oxidase/overoxidized polypyrrole bilayer based amperometric biosensor. *Bioelectrochemistry* **2018**, *124*, 47–56. [CrossRef]
51. Anibal, C.V.D.; Odena, M.; Ruisánchez, I.; Callao, M.P. Determining the adulteration of spices with Sudan I-II-II-IV dyes by UV–visible spectroscopy and multivariate classification techniques. *Talanta* **2009**, *79*, 887–892. [CrossRef] [PubMed]
52. Zainudin, N.S.; Yaacob, M.H.; Md Muslim, N.Z.; Othman, Z. Voltammetric determination of reactive black 5 (RB5) in waste water samples from the Batik industry. *Mal. J. Anal. Sci.* **2016**, *20*, 1254–1268.
53. Jäntschi, L.; Nașcu, H.-I. Chapter 4-Metode electrochimice. In *Chimie Analitică și Instrumentală*; Academic Press & Academic Direct: Cluj-Napoca, Romania, 2009; pp. 47–67.
54. Narayan, R.J. Part One-Fundamentals of medical biosensors for POC applications. In *Medical Biosensors for Point of Care (POC) Applications*; Woodhead Publishing: Sawston, UK, 2016; pp. 27–42.
55. Apetrei, I.; Apetrei, C. A modified nanostructured graphene-gold nanoparticle carbon screen-printed electrode for the sensitive voltammetric detection of rutin. *Measurement* **2018**, *114*, 37–43. [CrossRef]
56. Settle, F.A. Chapter 37-Voltammetric Techniques. In *Handbook of Instrumental Techniques for Analytical Chemistry*; Prentice Hall PTR: Upper Saddle River, NJ, USA, 1997; pp. 709–725.
57. Scholz, F. Voltammetric techniques of analysis: The essentials. *ChemTexts* **2015**, *1*, 17. [CrossRef]
58. Marsili, E.; Baron, D.B.; Shikhare, I.D.; Coursolle, D.; Gralnick, J.A.; Bond, D.R. Shewanella secretes flavins that mediate extracellular electron transfer. *Proc. Natl. Acad. Sci. USA* **2008**, *105*, 3968–3973. [CrossRef]
59. Hasanzadeh, M.; Shadjou, N.; Omidinia, E. Mesoporous silica (MCM-41)-Fe_2O_3 as a novel magnetic nanosensor for determination of trace amounts of amino acids. *Colloids Surf. B Biointerfaces* **2013**, *108*, 52–59. [CrossRef]
60. Saghatforoush, L.; Hasanzadeh, M.; Shadjou, N.; Khalilzadeh, B. Deposition of new thia-containing Schiff-base iron (III) complexes onto carbon nanotube modified glassy carbon electrodes as a biosensor for electrooxidation and determination of amino acids. *Electrochim. Acta* **2011**, *56*, 1051–1061. [CrossRef]
61. Hasanzadeh, M.; Karim-Nezhad, G.; Shadjou, N.; Hajjizadeh, M.; Khalilzadeh, B.; Saghatforoush, L.; Abnosi, M.H.; Babaei, A.; Ershad, S. Cobalt hydroxide nanoparticles modified glassy carbon electrode as a biosensor for electrooxidation and determination of some amino acids. *Anal. Biochem.* **2009**, *389*, 130–137. [CrossRef]
62. Hussain, M.M.; Rahman, M.M.; Asir, A.M. Sensitive L-leucine sensor based on a glassy carbon electrode modified with SrO nanorods. *Microchim Acta* **2016**, *183*, 3265–3273. [CrossRef]
63. Rezaei, B.; Zare, Z.M. Modified Glassy Carbon Electrode with Multiwall Carbon Nanotubes as a Voltammetric Sensor for Determination of Leucine in Biological and Pharmaceutical Samples. *Anal. Let.* **2008**, *41*, 2267–2286. [CrossRef]
64. Yang, D.X.; Zhu, L.D.; Jiang, X.Y. Electrochemical reaction mechanism and determination of Sudan I at a multi wall carbon nanotubes modified glassy carbon electrode. *J. Electroanal. Chem.* **2010**, *640*, 17–22. [CrossRef]
65. Li, J.; Kuang, D.; Feng, Y.; Zhang, F.; Xu, Z.; Liu, M.; Wang, D. Green synthesis of silver nanoparticles–graphene oxide nanocomposite and its application in electrochemical sensing oftryptophan. *Biosen. Bioelectr.* **2013**, *42*, 198–206. [CrossRef] [PubMed]
66. Barnes, W.L.; Dereux, A.; Ebbesen, T.W. Surface plasmon subwavelength optics. *Nat. Cell Biol.* **2003**, *424*, 824–830. [CrossRef] [PubMed]

67. Lai, G.; Zhang, H.; Yong, J.; Yu, A. In situ deposition of gold nanoparticles on polydopamine functionalized silica nanosphere for ultrasensitive nonenzymatic electrochemical immunoassay. *Biosen. Bioelectr.* **2013**, *47*, 178–183. [CrossRef]
68. García-Barrasa, J.; López-De-Luzuriaga, J.M.; Monge, M. Silver nanoparticles: Synthesis through chemical methods in solution and biomedical applications. *Cent. Eur. J. Chem.* **2011**, *9*, 7–19. [CrossRef]
69. Krishnaraj, C.; Jagan, E.G.; Rajasekar, S.; Selvakumar, P.; Kalaichelvan, P.T.; Mohan, N. Synthesis of silver nanoparticles using Acalypha indica leaf extracts and its antibacterial activity against water borne pathogens. *Colloids Surf. B Biointerfaces* **2010**, *76*, 50–56. [CrossRef]
70. Singha, S.; Saikia, J.P.; Buragohaina, A.K. A novel 'green' synthesis of colloidal silver nanoparticles (SNP) using Dillenia indica fruit extract. *Colloids Surf. B Biointerfaces* **2013**, *102*, 83–85. [CrossRef]
71. Gargi, D.; Dipankar, H.; Atanu, M. Synthesis of Gold Colloid using Zingiber officinale: Catalytic Study. *NanoMatChemBioDev* **2018**, *1*, 24–29.
72. Pulit, J.; Banach, M. Preparation of nanocrystalline silver using gelatin and glucose as stabilizing and reducing agents, respectively. *Dig. J. Nanomater. Biostruct.* **2013**, *8*, 787–795.
73. Prabakaran, E.; Pandian, K. Amperometric detection of Sudan I in red chili powder samples using Ag nanoparticles decorated graphene oxide modified glassy carbon electrode. *Food Chem.* **2015**, *166*, 198–205. [CrossRef]
74. Liu, X.; Luo, L.; Ding, Y.; Kang, Z.; Ye, D. Simultaneous determination of L-cysteine and L-tyrosine using Au-nanoparticles/poly-eriochrome black T film modified glassy carbon electrode. *Bioelectrochemistry* **2012**, *86*, 38–45. [CrossRef] [PubMed]
75. Prabakaran, E.; Sheela Violet Rani, V.; Brabakaran, A.; Pandian, K.; Jesudurai, D. A Green Approach to the Synthesis of Eriochrome Black-T Capped Silver Nanoparticles and Its Electrochemical Detection of L-Tryptophan and L-Tyrosine in Blood Sample and Antibacterial Activity. *J. Adv. Electrochem.* **2016**, *2*, 78–84.
76. Karastogianni, S.; Girousi, S. A novel electrochemical bioimprinted sensor for butyl paraben on a modified carbon paste electrode with safranine-O capped with silver nanoparticles. *Int. J. Cur. Res.* **2017**, *9*, 61118–61124.
77. García-Carmona, L.; González, M.C.; Escarpa, A. Electrochemical On-site Amino Acids Detection of Maple Syrup Urine Disease Using Vertically Aligned Nickel Nanowires. *Electroanalysis* **2018**, *30*, 1505–1510. [CrossRef]
78. Tooley, C.A.; Gasperoni, C.H.; Marnoto, S.; Halpern, J.M. Evaluation of Metal Oxide Surface Catalysts for the Electrochemical Activation of Amino Acids. *Sensors* **2018**, *18*, 3144. [CrossRef]
79. Nguyen, H.H.; Lee, S.H.; Lee, U.J.; Fermin, C.D.; Kim, M. Immobilized Enzymes in Biosensor Applications. *Materials* **2019**, *12*, 121. [CrossRef]
80. Clark, L.C.; Lyons, C. Electrode systems for continuous monitoring in cardiovascular surgery. *Ann. N. Y. Acad. Sci.* **1962**, *102*, 29–45. [CrossRef]
81. García-Carmona, L.; González, M.C.; Escarpa, A. On-line coupling of millimeter size motors and chronoamperometry for real time bio-sensing of branched-chain amino acids in maple syrup urine disease clinical samples. *Sens. Actuators B Chem.* **2019**, *281*, 239–244. [CrossRef]
82. Stefan-van Staden, R.-I.; Muvhulawa, L.S. Determination of L- and D-Enantiomers of Leucine Using Amperometric Biosensors Based on Diamond Paste. *Instr. Sci. Technol.* **2006**, *34*, 475–481. [CrossRef]
83. Labroo, P.; Cui, Y. Amperometric bienzyme screen-printed biosensor for the determination of leucine. *Anal. Bioanal. Chem.* **2014**, *406*, 367–372. [CrossRef]
84. Domınguez, R.; Serra, B.; Reviejo, A.; Pingarron, J. Chiral analysis of amino acids using electrochemical composite bienzyme biosensors. *Anal. Biochem.* **2001**, *298*, 275–282.
85. Cosnier, S.; Lepellec, A. Biosensors based on electropolymerized films: New trends. *Anal. Bioanal. Chem.* **2003**, *377*, 507–520. [CrossRef] [PubMed]
86. Garnier, F. Functionalized Conducting Polymers—Towards Intelligent Materials. *Angew. Chem.* **1989**, *101*, 529–533. [CrossRef]
87. Luo, J.; Fan, C.; Wang, X.; Liu, R.; Liu, X. A novel electrochemical sensor for paracetamol based on molecularly imprinted polymeric micelles. *Sens. Act. B Chem.* **2013**, *188*, 909–916. [CrossRef]
88. Kan, X.; Zhou, H.; Li, C.; Zhu, A.; Xing, Z.; Zhao, Z. Imprinted electrochemical sensor for dopamine recognition and determination based on a carbon nanotube/polypyrrole film. *Electrochim. Acta* **2012**, *63*, 69–75. [CrossRef]

89. Ghasemi-Varnamkhasti, M.; Apetrei, C.; Lozano, J.; Anyogu, A. Potential use of electronic noses, electronic tongues and biosensors as multisensor systems for spoilage examination in foods. *Trends Food Sci. Technol.* **2018**, *80*, 71–92. [CrossRef]
90. Apetrei, I.-M.; Apetrei, C. Application of voltammetric e-tongue for the detection of ammonia and putrescine in beef products. *Sens. Actuators B Chem.* **2016**, *234*, 371–379. [CrossRef]
91. Wang, Z.; Chen, H.; Li, J.; Xue, Z.; Wu, B.; Lu, X. Acetylsalicylic acid electrochemical sensor based on PATP–AuNPs modified molecularly imprinted polymer film. *Talanta* **2011**, *85*, 1672–1679. [CrossRef]
92. Bi, Q.; Dong, S.; Sun, Y.; Lu, X.; Zhao, L. An electrochemical sensor based on cellulose nanocrystal for the enantioselective discrimination of chiral amino acids. *Anal. Biochem.* **2016**, *508*, 50–57. [CrossRef]
93. Ríos, A.; Zougagh, M.; Avila, M. Miniaturization through lab-on-a-chip: Utopia or reality for routine laboratories? A review. *Anal. Chim. Acta* **2012**, *740*, 1–11. [CrossRef]
94. Batalla, P.; Martín, A.; López, M.A.; González, M.C.; Escarpa, A. Enzyme-based microfluidic chip coupled to graphene electrodes for the detection of D-amino acid enantiomer-biomarkers. *Anal. Chem.* **2015**, *87*, 5074–5078. [CrossRef] [PubMed]

© 2020 by the authors. Licensee MDPI, Basel, Switzerland. This article is an open access article distributed under the terms and conditions of the Creative Commons Attribution (CC BY) license (http://creativecommons.org/licenses/by/4.0/).

MDPI
St. Alban-Anlage 66
4052 Basel
Switzerland
Tel. +41 61 683 77 34
Fax +41 61 302 89 18
www.mdpi.com

Applied Sciences Editorial Office
E-mail: applsci@mdpi.com
www.mdpi.com/journal/applsci

www.ingramcontent.com/pod-product-compliance
Lightning Source LLC
LaVergne TN
LVHW070543100526
838202LV00012B/366